THE DISTRACTED MIND

THE DISTRACTED MIND

Ancient Brains in a High-Tech World

ADAM GAZZALEY
AND LARRY D. ROSEN

The MIT Press
Cambridge, Massachusetts
London, England

This book was set in Adobe Garamond Pro and Berthold Akzidenz Grotesk by the MIT Press. Printed and bound in the United States of America.

Library of Congress Cataloging-in-Publication Data

Names: Gazzaley, Adam, author. | Rosen, Larry D., author.
Title: The distracted mind : ancient brains in a high-tech world / Adam
    Gazzaley and Larry D. Rosen.
Description: Cambridge, MA : MIT Press, [2016] | Includes bibliographical
    references and index.
Identifiers: LCCN 2016010271 | ISBN 9780262034944 (hardcover : alk. paper)
Subjects: LCSH: Distraction (Psychology) | Attention. | Information society.
    | Information technology--Social aspects.
Classification: LCC BF323.D5 G39 2016 | DDC 153.7/3--dc23 LC record available at
https://lccn.loc.gov/2016010271

10 9 8 7 6 5 4 3 2

Contents

Acknowledgments

THIS STORY OF *The Distracted Mind* began for me in 2002 when I moved to San Francisco from the East Coast for a postdoctoral fellowship in cognitive neuroscience at UC Berkeley. I was searching for a research project that was not driven solely by the goal of advancing psychological theory, but rather one that everyday people would care about because its results would teach them something about their minds that truly mattered to them. And so, inspired by repeatedly hearing descriptions of the negative impact of distraction on the lives of my older patients at the neurology clinic at UC San Francisco, I set out to study the impact of distraction on memory.

- I would like to start by acknowledging those patients who were trusting enough to allow me to peer into their lives as they shared with me their insecurities and the fragilities of their minds. They were the first to open my eyes to the reality of the Distracted Mind.
- I am grateful to my first scientific mentor, John Morrison, and my mentors at UC Berkeley who gave me the opportunity, encouragement, and intellectual support to launch this research program: Mark D'Esposito and Robert Knight.
- I thank my lab mates who helped me develop and test the paradigms we used to assess the neural mechanisms of distractibility, which are referenced throughout the book: Brian Miller, Jesse Rissman, Jeff Cooney, Aaron Rutman, Kevin McEvoy, Tyler Seibert, Jon Kelley, and Darya Pino (now Darya Rose). Darya deserves special thanks for over a decade of enlightening discussions on this topic.
- I am indebted to the team of faculty, postdocs, research associates, and students from my lab, who worked tirelessly over many years to elucidate the details of

what occurs in our brains to generate the Distracted Mind, and how we might remediate its negative impact: Theodore Zanto, Wes Clapp, Zack Chadick, Michael Rubens, Jacob Bollinger, Jonathan Kalkstein, Jyoti Mishra, Joaquin Anguera, Ezequiel Morsella, Anne Berry, Peter Wais, Brad Voytek, Nate Cashdollar, Cammie Rolle, Judy Pa, David Ziegler, WanYu Hsu, Omar Al-Hashimi, Jacki Janowich, Jean Rintoul, and Jaqueline Boccanfuso. Sincere thanks to the many other members of our lab who assisted on these projects, the volunteer interns who made all of this research possible, and of course the hundreds of research participants over the years who gave so much time and effort to advance our science.

The next phase of the story began in 2009, when I was invited to give a talk at the Annual AARP meeting. This was the first presentation of my research on distraction that opened my eyes to the power of sharing science with the public. It led to hundreds of talks around the world over the subsequent years and my PBS special, *The Distracted Mind with Adam Gazzaley.*

- I acknowledge all of those audience members who asked so many wonderful, heartfelt questions, which drove me to search deeper for answers to the issues that people really care about.
- I thank Lennlee Keep and the team from Santé Fe Productions who encouraged and guided me through the intense process of creating the PBS show. That journey helped me distill my thoughts on this topic and learn how to best share it with the public.

My story of the Distracted Mind culminated with the writing of this book. The truth is I really did not want to write a book at all; I spend so much of time in scientific writing and I most enjoy the stage and film when it comes to sharing with the public.

- I would not have written this book if not for Larry Rosen. His enthusiasm for this project and the valuable insights he brought to the collaboration in terms of the impact of technology on the Distracted Mind convinced me that this story needed to be told with the depth and breadth that can only be achieved with a book.
- I thank our agent Stacey Glick, who has always had our back; Bob Prior, at the MIT Press, who believed in this book from the very beginning; Cammie Rolle,

who helped with literature research; and our copy editor, Judith Feldmann, who put up with all of our pickiness.

Finally, a deep thank you to my wife and love, Jo Gazzaley, who contributed tireless attention to refining the details of this book. For her selfless devotion to working by my side to elevate this book to the next level, I am eternally grateful.

—AG

A FEW YEARS AGO, after publishing my fifth book on the psychological impact of technology, I wondered whether there was anything left for me to say in the long format of a book rather than in my shorter blog posts. I had written about how technology was starting to make us crazy in *TechnoStress* (way back in 1997!), about parenting high-tech children in *Me, MySpace, and I* (2007), and about educating a new generation of tech-savvy students in *Rewired* (2010). In 2012 I wrote about how we were all becoming obsessed with our technology in *iDisorder* and in 2015 I coedited the Wiley-Blackwell handbook, *Psychology, Technology, and Society*. As I immersed myself in my lab's psychological research I realized that to fully understand the impact of technology I had to expand my perspective to the field of neuroscience. Serendipitously, after delivering the opening keynote at the Learning and the Brain Conference in San Francisco, I attended a session on multitasking and was blown away by Dr. Adam Gazzaley's carefully crafted neuroscience research. I rarely take notes at conferences and found myself with several pages of results and diagrams followed by immersing myself into the groundbreaking work coming out of his lab, including being spellbound by his video lecture on the Distracted Mind. Out of the blue I sent Adam an email and after a few back-and-forth emails and a couple of long phone calls we decided to write this book together. I was overjoyed. This book combines our two specialty areas—neuroscience and psychology—in a unique look at how and why we have become such a distracted society.

I am deeply indebted to Dr. Mark Carrier and Dr. Nancy Cheever—my cofounders of the George Marsh Applied Cognition Lab at California State University, Dominguez Hills—as well as all the undergraduate and graduate students who have worked on research in our lab. By my count in the six years since we combined our research into one lab, we have minted two PhDs and currently have fifteen GMAC Lab grads in PhD programs, along with others having completed their MA degrees and working in the field. Special thank go to Alex Lim, MA, who is our fourth lab "mentor" and tirelessly runs our new neuroscience lab.

I would be remiss if I did not acknowledge the role that California State University, Dominguez Hills, has played in the last forty-one years of my life. CSUDH is a small state university and many of our students are the first in their family to attend college, let alone aspire for a higher degree. The campus has supported my work by allowing me to teach "The Global Impact of Technology" for the last ten years—the only class taught in the university theater—to 450 students each semester. They have furthered my work by providing me with eager students and by funding our research with small grants here and there to keep our lab running. Federal programs support many of our students, and we thank the McNair Scholars Program, MBRS-RISE, and MARC-USTAR, for their support. President Willie Hagan deserves a special thank you for funding our neuroscience lab after hearing me speak to several campus groups about the coming need for neuropsychological research to complement and expand our lab focus on the "psychology of technology."

On a personal level I could not complete any book without my fiancée Dr. Vicki Nevins who has supported me through the last five books even though I kept saying that each book was my last. She listened and applauded me when I was high from a great writing day, and commiserated with me when I was frustrated on less productive days. My four children and their spouses are all amazing. Adam and Farris, Arielle and Jess, Chris and Tiffany, and Kaylee and Grant take time to make sure we share our lives even though none of them live close enough to see face to face all that often. We text, FaceTime, Facebook, and actually (gasp) talk on the phone, and I feel their love and caring with me 24/7. Three special hugs go to Grayson (my granddaughter) and Evan and Michael (Vicki's grandsons, who have always called me grandpa even though I am technically not yet "official").

A big thank you goes to Bob Prior at the MIT Press for believing in this book. Judith Feldmann did a wonderful copyediting job amid our cutting and pasting and moving and adding. Adam's wife, Jo Gazzaley, played a major role in reading the entire manuscript many times and finding areas for enhancement, correction, and improvement. Last, but certainly not least, Stacey Glick has been my agent for my last five books and has always been available to help me understand the publishing world.

—LDR

Prologue

THIS BOOK IS the first of its kind to explore the daily challenges we face with the highly engaging but extremely distracting high-tech world we now inhabit, from the dual points of view of a psychologist and a neuroscientist. By providing both scientific foundations and real-world examples of people facing and addressing their own distracted minds, we share with you a unique perspective on how our increasingly information-saturated world (overflowing with pop-up windows, smartphones, texts, chat, email, social media, and video games) has coupled with growing expectations of 24/7/365 availability and immediate responsiveness to place excessive demands on our brains. *The Distracted Mind* will take you on a journey into how and why we struggle with interruptions and distractions that emerge from both our inner and outer worlds, as well as offer practical strategies for changing your behavior and enhancing your brain function to alleviate interference and better accomplish your goals. It is clear that our interruptive technologies are only going to become more effective in drawing our attention away from important aspects of life, so we urgently need to understand why we are so sensitive to interference and how we can find a "signal amidst the noise" in our high-tech world.

The Distracted Mind is not a pseudo-science book that offers colorful brain scans and questionable neuroscience as a way of making a topic appear more credible. In this book, we apply our complementary scientific lenses to present timely and practical insights. Dr. Adam Gazzaley is a cognitive neuroscientist and a trailblazer in the study of how the brain manages distractions and interruptions. Dr. Larry Rosen is a psychologist who has studied the "psychology of technology" as a pioneer in this field for more than thirty years. Our complementary perspectives focus on demonstrating why we fail to successfully navigate our modern technological ecosystem and how that has detrimentally affected our safety, cognition, education, workplace, and our relationships with family and friends. We enrich this discussion with our own research and scientific hypotheses, as well as views of other scholars in the field, to explain why our brains struggle to keep up with demands of communication and information.

We present our perspectives in three parts. In Part I, we will take you on a tour through new insights into why our "interference dilemma" exists in the first place and why it has become so relevant to us now. We describe how the very essence of what has evolved furthest in our brains to make us human—our ability to set high-level goals for ourselves—collides head-first with our brain's fundamental limitations in cognitive control: attention, working memory, and goal management. This collision results in our extreme sensitivity to goal interference from both distractions by irrelevant information and interruptions by attempted multitasking. This noise degrades our perceptions, influences our language, hinders effective decision making, and derails our ability to capture and recall detailed memories of life events. The negative impact is even greater for those of us with undeveloped or impaired cognitive control, such as children, teens, and older adults as well as many clinical populations. We further discuss why we engage in high-interference-inducing behaviors from an evolutionary perspective, such that we are merely acting in an optimal manner to satisfy our innate drive as information-seeking creatures.

In Part II, we will share a careful assessment of our real-world behaviors and demonstrate how the collision described in Part I has been intensified by our constant immersion with the rich landscape of modern information technology. People do not sit and enjoy a meal with friends and family without checking their phones constantly. We no longer stand idle in waiting lines, immersed in thought or interacting with those next to us. Instead, we stare face down into virtual worlds beckoning us through our smartphones. We find ourselves dividing our limited attention across complex demands that often deserve sustained, singular focus and deeper thought. We will share our views of why we behave in such a manner, even if we are aware of its detrimental effects. Building a new model inspired by *optimal foraging theory* we explain how our high-tech world perpetuates this behavior by offering us greater accessibility to feed our instinctive drive for information as well as influencing powerful internal factors, such as boredom and anxiety. We are most certainly *ancient brains in a high-tech world*.

Finally, in Part III we offer our perspectives on how we can change our brains to make them more resilient, as well as how we can change our behavior via strategies to allow us to thrive in all areas of our lives. We first explore the full landscape of potential approaches available to us—from the low-tech to the high-tech—that harness our brain's plasticity to strengthen our Distracted Mind. This in-depth examination includes traditional education, cognitive training, video games, pharmaceuticals, physical exercise, meditation, nature exposure, neurofeedback, and brain stimulation, illustrating how in these fascinating times the same technologies that aggravate the Distracted Mind can be flipped around to offer remediation. We then share advice on what we can do from a strategic perspective to modify our behavior, without abandoning modern technology, such that we minimize the negative consequences of having a Distracted Mind. Using the optimal foraging model introduced earlier in the book as a framework to approach behavioral change, all of the strategies we offer are practical and backed by solid science.

The Distracted Mind will enlighten you as to how and why our brains struggle to manage a constantly surging river of information in a world of unending interruptions and enticements to switch our focus. We extend this perspective to explore the consequences of this overload on our ability to function successfully in our personal lives, on the road, in classrooms, and in the workplace, and address why it is that we behave in this way. Critically, we provide solid, down-to-earth advice on what we need to do in order to survive and thrive in the information age.

Part I

COGNITION AND THE ESSENCE OF CONTROL

YOUR BRAIN IS an incredible information-processing system and the most complex structure known to humankind. The brain has allowed us to perform extraordinary feats from discovering general relativity to painting the Sistine Chapel, from building airplanes to composing symphonies. And yet, we still forget to pick up milk on the way home. How can this be?

In this first part of the book, we explain how the collision between our goals and our limited cognitive control abilities leads to interference and diminished performance. In chapter 1, we begin by discussing interference in detail: what it is, how it affects us, and why it seems to be getting worse. Here we propose that a model originally created to explain foraging behavior in animals can be adapted to understand our propensity for task switching. In chapter 2, we look at how the human brain has evolved to establish complex goals and implement them using a set of abilities known as cognitive control: attention, working memory, and goal management. Chapter 3 is a deep dive into the brain. The development of noninvasive technologies over the last couple decades has allowed us to peer into a functioning human brain, enabling us to much better understand brain processes and cognitive control. As impressive as these abilities are, they are also subject to fundamental limitations that have not changed all that much from our ancient ancestors. In chapter 4, we explore these limitations in our attention, working memory, and goal management. Finally, in chapter 5, we examine how these limitations are affected by age, clinical conditions, and even day-to-day changes in our internal state.

1 INTERFERENCE

WE HAVE COME to believe that the human brain is a master navigator of the river of information that rages steadily all around us. And yet we often feel challenged when trying to fulfill even fairly simple goals. This is the result of *interference*—both *distractions* from irrelevant information and *interruptions* by our attempts to simultaneously pursue multiple goals. Many of you may now be glancing accusingly at your mobile phone. But before we place any blame on this potential culprit, it is critical to understand that our sensitivity to interference, or what we will refer to throughout this book as "the Distracted Mind," was not born out of modern technology. Rather, it is a fundamental vulnerability of our brain. Consider these three scenarios, which could just as easily have happened to you today as over a hundred years ago:

- You step into your kitchen and open the refrigerator (or icebox), and your mood sinks as you realize that you have absolutely no recollection of what it was you wanted to retrieve. How is this possible? Surely you can remember a single item for the several seconds it took you to arrive there. A bit of introspection and you realize that this was not the result of a pure "memory" glitch, but rather the result of interference—you were distracted from your goal by intrusive thoughts of an upcoming meeting.
- You sit at a meeting, staring at your colleague across the table in a crowded restaurant (or watering hole), struggling to follow her story. You can hear her just fine, but it seems that your brain keeps getting hijacked by the chatter of the room around you, even though you are desperately trying to ignore these distractions.

- You walk home from the meeting through an unfamiliar part of town, but instead of focusing on your route you keep thinking about your conversation; the next thing you know you have lost your way. Interruption generated by your own mind derailed you from successfully accomplishing your goal.

Despite our brain's inherent sensitivity to interference, it is undeniable that recent technological advances have made things more difficult for the Distracted Mind. Welcome to our new reality:

- You are at a meeting, and although an important new project is being discussed and the use of devices during meetings is seriously frowned upon, you sneak a peek at the mobile phone in your lap to check the status of an email you were expecting … and while you are at it, you casually scan a few social media sites to catch up on your friends' latest activities.
- You are at the dinner table with your family, the television blaring in the background, and everyone has a phone sitting on the table that they constantly pick up, check the screen, tap a few times, and then put down face up so they don't miss incoming messages. This is followed by clumsy attempts to figure out what you missed in the conversation and reengage as best as possible.
- You are cruising along the highway at 60 mph. You feel that familiar buzz in your pocket—you got a text! You certainly know better, but you reach for your phone anyway, guiltily glancing over at the driver in the car next to you.
- Your child uses an iPad as part of a new school program to introduce technology into the classroom. Seems like a good idea until you get a call from your son's teacher who informs you that that he is not using it for the intended purpose; instead, he is constantly playing video games and downloading apps during class.
- You plop down at your desk with a heavy burden on your mind of an important assignment due by the end of the day. Despite the fact that high-level performance on this task is critical for your job evaluation, you find yourself constantly checking your email and Facebook. As the day goes by, each interruption sets off a chain reaction of communication that increases your chances of missing your deadline. You know you need to focus on your task, yet you continue down this doomed path.

So many technological innovations have enhanced our lives in countless ways, but they also threaten to overwhelm our brain's goal-directed functioning with interference. This interference has a detrimental impact on our cognition and behavior in daily activities. It impacts every level of our thinking, from our perceptions, decision making, communication, emotional regulation, and our memories. This in turn translates into negative consequences on our safety, our education, and our ability to engage successfully and happily with family, friends, and colleagues. The magnitude of the impact is even greater for those of us with underdeveloped or impaired brains, such as children, older adults, and individuals suffering from neurological and psychiatric conditions. If we hope to successfully manage interference, we first must understand its nature.

WHAT IS GOAL INTERFERENCE?

"Interference" is a general term used to describe something that hinders, obstructs, impedes, or largely derails another process. When you pick up static on a radio station, you are detecting interference with the reception of radio waves that are relevant to you; this is also referred to as "noise." The goal interference described in the above scenarios is in many ways not very different from radio noise. This type of interference has been the focus of extensive research by a diverse group of experts in fields as varied as psychology, neuroscience, education, advertising, marketing, and human factors; but it is not often presented as a unified construct, which is one of the main goals of this book.

Goal interference occurs when you reach a decision to accomplish a specific goal (e.g., retrieve something from the refrigerator, complete a work assignment, engage in a conversation, drive your car) and something takes place to hinder the successful completion of that goal. The interference can either be generated *internally*, presenting as thoughts within your mind, or generated *externally*, by sensory stimuli such as restaurant chatter, beeps, vibrations, or flashing visual displays (figure 1.1). Goal interference, originating from either your internal or external environments (often both), can occur in two distinct varieties—distractions and interruptions—based on your decision about how you manage the interference.[1]

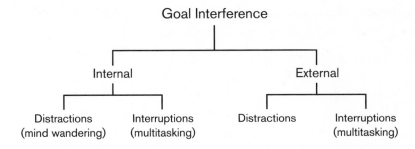

Figure 1.1
A conceptual framework of goal interference that can be generated both internally and externally, and can be elicited by both goal-irrelevant information (distractions) and multitasking (interruptions).

To understand these different types of goal interference, we find it helpful to temporarily set aside a discussion of the influence of technology and consider a scenario that has taken place for millennia—the act of sitting down with a friend to catch up on each other's lives. This seems like a relatively straightforward goal. But, even without the presence of modern technology, four types of interference threaten to derail you from accomplishing this sort of goal: internal distraction, external distraction, internal interruption, and external interruption. Let's dive in and discuss each of them.

Distractions are pieces of goal-irrelevant information that we either encounter in our external surroundings or generate internally within our own minds. When it comes to distractions, our intentions are very clear— we wish to ignore them, shut them out, suppress them, and carry on with attempting to accomplish our singular goal. Consider the following sort of very common situation:

You are engaged in that interesting conversation with your friend, but then your mind wanders against your will to something completely irrelevant to the conversation: "I can't believe my boss didn't notice how much I accomplished this week!"

This is an example of an internal distraction, also sometimes known as mind wandering. Mind wandering is frequently negative in content, as in this scenario.[2] Just as often, however, distractions are generated externally by surrounding sights, sounds, and smells that are irrelevant to your goals, as in the following situation.

You are listening to your friend when you hear your name mentioned at a table nearby. Even though you already heard it before and are sure that they are not referring to you, hearing your own name captures your attention against your will and shifts your focus away from your goal.

And so, in a manner similar to what occurs when your mind wanders, information that is irrelevant to your goals can result in interference that we refer to as external distractions. Even if it is clear to you that external distractions will derail your conversation—and you are resolved to ignore them—they still often penetrate your mind and divert your attention away from your goals, thus diminishing your performance.

Interruptions are the other major source of goal interference. The difference from distractions is that interruptions happen when you make a decision to concurrently engage in more than one task at the same time, and even if you attempt to switch rapidly between them. Just like distractions, interruptions may be generated internally or externally. To appreciate an internally generated interruption, let's revisit the conversation you were having with your friend.

The conversation has become much less interesting to you. And so, you decide to fragment off a bit of your focus and direct it toward thinking about how your boss perceives your efforts at work, all the while attempting to maintain the conversation with your friend.

This act of voluntarily engaging in a concurrent, secondary internal task is an internally generated interruption. It generates interference by derailing your goal of having a meaningful conversation. Interruptions are also frequently generated externally.

Now while engaged in conversation with your friend, you overhear a fascinating conversation taking place nearby and make the decision to simultaneously eavesdrop as you continue your own conversation.

Interruptions such as these are often referred to as "multitasking," defined as the act of attempting to engage simultaneously in two or more tasks that have independent goals. The word "attempting" is used here because, as you will see later in the book, multitasking may be the behavior you decide to engage in, but when it comes to what actually occurs in your brain, the term "task switching" is a better description.

Interestingly, the actual content of goal interference can be the same for distractions and interruptions. In our example, thoughts of your boss's impression of your work quality were the source of interference for both internal distraction and interruption, and an overheard conversation was the source of both external distraction and interruption. What distinguishes distractions from interruptions are your intentions about how you choose to manage them; either you attempt to ignore them and carry on with your original goal—distraction—or you engage in them as a simultaneous, secondary goal—interruption. Despite both of these being types of goal interference, different brain mechanisms underlie the performance impairment they generate, as we will discuss later.

WHY ARE WE SO SUSCEPTIBLE TO INTERFERENCE?

All complex systems are susceptible to interference, including the functioning of our cars, laptops, 747s, and the Hubble Telescope. The opportunity for interference to degrade any system's performance seems to scale with its complexity. When it comes to the human brain, undeniably the most complex system in the known universe, it should thus come as no surprise that it is extremely sensitive to interference at many levels. Indeed, the reason why goal interference in particular is so prominent in our lives is the inherent complexity of our goals and the limitations we have in fulfilling them. Our ability to establish high-level goals is arguably the pinnacle of human brain evolution.[3] Complex, interwoven, time-delayed, and often shared goals are what allow us humans to exert an unprecedented influence over how we interact with the world around us, navigating its multifaceted environments based on our decisions rather than reflexive responses to our surroundings. Our impressive goal-setting abilities have permitted the remarkable development of our cultures, communities, and societies and have enabled us to create complex human constructs, such as art, language, music, and technology.

The sheer magnitude of our impressive goal-setting abilities has resulted in the conditions necessary for goal interference to exist in the first place.

Our proficiency in setting goals is mediated by a collection of cognitive abilities that are widely known as "executive functions," a set of skills that include evaluation, decision making, organization, and planning. But goal setting is only half the battle. We also need specialized processes to *enact* all those lofty goals. Our ability to effectively carry out our goals is dependent on an assemblage of related cognitive abilities that we will refer to throughout this book as "cognitive control." This includes attention, working memory, and goal management. Note that our ability to set high-level goals does not necessarily mean that it is inevitable that we are overwhelmed by goal interference. It is conceivable that the goal-enactment abilities of our brain evolved alongside our goal-setting abilities to offset any negative impact of goal interference. But this is not what seems to have happened. Our cognitive control abilities that are necessary for the enactment of our goals have not evolved to the same degree as the executive functions required for goal setting. Indeed, the fundamental limitations in our cognitive control abilities do not differ greatly from those observed in other primates, with whom we shared common ancestors tens of millions of years ago.[4]

Our cognitive control is really quite limited: we have a restricted ability to distribute, divide, and sustain attention; actively hold detailed information in mind; and concurrently manage or even rapidly switch between competing goals. We can only speculate that if the neural processes of goal enactment evolved to a comparable degree as our goal-setting abilities, we would not be so encumbered by goal interference. If we could hold more information in mind and with higher fidelity, if we could cast a broader attentional net on the world around us and with greater sustainability, if we could simultaneously engage in multiple demanding tasks and transition more efficiently between them, we would not be so easily distracted and interrupted. *In many ways, we are ancient brains in a high-tech world.*

We can visualize this as a conflict between a mighty force, represented by our goals, which collides head on with a powerful barrier, represented by the limitations to our cognitive control. The conflict is between our goal-setting abilities, which are so highly evolved, driving us to interact in high-interference environments to accomplish our goals, and our goal-enactment abilities, which have not evolved much at all from our primitive ancestors, representing fundamental limitations in our ability to process information. It is this conflict that results in goal interference, and generates a palpable tension in our minds—*a tension between what we want to do and what we can do.* Your awareness of this conflict, even if only at a subconscious level, is likely what led you to pick up this book in the first place. That, and a dawning realization that this conflict is escalating into a full-scale war, as modern technological advancements worsen goal interference to further besiege the Distracted Mind.

IS IT GETTING WORSE?

Humans have always lived in a complex world, one rich with enticing distractions and teeming with countless interruptions via alternative activities that threaten to bar us from accomplishing our goals. While goal interference has likely existed for as long as modern humans have walked the Earth, the last several decades have witnessed profound changes: The Information Age has emerged on the heels of modern technological breakthroughs in computers, media, and communication. This latest stage in human history may have been sparked by the digital revolution, but the rise of personal computers, the Internet, smartphones, and tablets is really only the surface. The true core of the change to our mental landscape is that we are experiencing an elevation of information itself to the level of the ultimate commodity. This has fueled an ever-expanding explosion in the variety and accessibility of technologies with enticing sounds, compelling visuals, and insistent vibrations that tug at our attention while our brains attempt to juggle multiple streams of competing information.

Most of us now carry a small device on us that is as powerful, if not more so, than the computers that sat on our desks only a decade ago. Smartphones are quickly becoming ubiquitous. According to a 2015 report by the Pew Research Center, 96 percent of all US adults own a mobile phone, and 68 percent own a smartphone. Among US smartphone users, 97 percent of them regularly use their phone to send text messages, 89 percent use it to access the Internet, and 88 percent send and receive email.[5] Worldwide estimates are that 3.2 billion people, 45 percent of the world's population, own a mobile phone.[6] Beyond this evidence of global penetration and the fact that these devices now dwell in our pockets and purses, new media also facilitate constant switching. Smartphones, desktops, and laptops support multiple apps while web browsers allow numerous simultaneously open tabs and windows, making it increasingly difficult to attend to a single website or app without having our attention lured away. This new pattern of engagement extends to the way that we use different types of media. There is a well-documented and growing tendency for many of us to "media multitask." For example, a study by Dr. Rosen's lab found that the typical teen and young adult believes that he or she can juggle six to seven different forms of media at the same time.[7] Other studies have shown that up to 95 percent of the population report media multitasking each day, with activity in more than one domain occupying approximately a third of the day.[8]

Moreover, these technological innovations have been accompanied by a shift in societal expectations such that we now demand immediate responsiveness and continuous productivity. Several studies have reported that US adults and teenagers check their phone up to 150 times a day, or every six to seven minutes that they are awake.[9] Similar studies in the UK have found that more than half of all adults and two-thirds of young adults and teens do not go one hour without checking their phones. Furthermore, three in four smartphone owners in the US feel panicked when they cannot immediately locate their phone, half check it first thing in the morning while still lying in bed, one in three check it while using the bathroom, and three in ten check it while dining with others. According to a Harris Poll, eight in ten vacationers brought or planned to bring at least one high-tech device on vacation, and a substantial proportion of vacationers checked in often with their devices.[10]

Constant accessibility, invasive notifications, task-switching facilitators, and widespread changes in expectations have escalated and perpetuated our interference dilemma. Indeed, it seems likely that these wonders of our modern technological world have generated a higher level of goal interference than we have ever experienced. Although this societal trend taxes our fragile cognitive control abilities to a breaking point for some of us, it nevertheless persists, and by all indications is escalating rapidly. While from some perspectives this may be considered a more enlightened time, our behavior in this domain seems to be completely incongruent with the very nature of our pursuit of our goals—something that is fundamental to our very humanity.

WHY DO WE BEHAVE THIS WAY?

Despite an emerging awareness of our sensitivity to goal interference and the widespread negative impact it can have on our lives, most of us engage in interference-inducing behaviors, even when distractions and multitasking may be completely avoidable. Interference-inducing behaviors involve intentionally placing oneself in a distracting environment (e.g., going to a crowded, noisy coffee shop to write a book) or engaging in multitasking behavior (e.g., writing a book while listening to music and routinely checking incoming texts and email). Almost no one seems to be immune to engaging in these behaviors. And so a fascinating question remains—why do we do it, even if we understand that it degrades our performance?

A common explanation that is offered in response to this question is that it is simply more fun and rewarding to engage in multitasking compared to single tasking. There certainly appears to be truth in this assertion. Individuals report that enjoyment is a factor in Internet-based multitasking and that performing additional tasks while watching TV ads increases overall task enjoyment.[11] Also in support of this view, physiological signs of increased arousal are associated with switches between multiple types of content on a single device.[12] In regard to rewards, researchers have shown that novelty is associated with reward processing in our brains.[13] This is not surprising, as novelty seeking is a powerful driving force to explore new environments, and thus offers clear survival advantages. The novelty load is undoubtedly higher when frequently switching between new tasks than when just staying put, so it is logical that the overall reward gains, and thus the fun factor as well, are heightened when multitasking. In addition, the act of receiving an earlier reward is often more highly valued, even if a delayed reward has greater overall associated value.[14] This phenomenon, known as the "temporal discounting of rewards," is a strong influence on impulsive behaviors and so may also play a role in the inherent drive to seek the immediate gratification that comes from switching to new tasks sooner rather than later.

But we have always had plenty of opportunities to rapidly switch to novel, and thus more rewarding, alternative tasks. It seems that there is something more going on here than general reward seeking and fun. What is it about the modern technological world that has resulted in this frenzied multitasking behavior? In this book we will explore a novel hypothesis: We engage in interference-inducing behaviors because, from an evolutionary perspective, we are merely acting in an optimal manner to satisfy our innate drive to seek *information*. Critically, the current conditions of our modern, high-tech world perpetuate this behavior by offering us greater accessibility to feed this instinctive drive and also via their influence on internal factors such as boredom and anxiety.

How can self-perpetuated interference-inducing behaviors be considered optimal from any perspective, when they are clearly detrimental to us in so many ways? The answer is that at our core we are *information-seeking creatures*, so behaviors that maximize information accumulation are optimal, at least from that viewpoint. This notion is supported by findings that molecular and physiological mechanisms that originally developed in our brain to support food foraging for survival have now evolved in primates to include information foraging.[15] Data to support this assertion rest largely on observations that the dopaminergic system, which is crucial for all reward processing, plays a key role in both basic food-foraging behavior in lower vertebrates and higher-order cognitive behaviors in monkeys and humans that are often dissociated from clear survival benefits.[16] The role of the dopamine system has actually been shown to relate directly to information-seeking behavior in primates. Macaque monkeys, for example, respond to receiving information similarly to the way they respond to primitive rewards such as food or water. Moreover, "single dopamine neurons process both primitive and cognitive rewards, and suggest that current theories of reward-seeking must be revised to include information-seeking."[17]

As Thomas Hills, a pioneer of this perspective, describes, "Evidence strongly supports the evolution of goal-directed cognition out of mechanisms initially in control of spatial foraging but, through increasing cortical connections, eventually used to forage for information."[18] The claim that we are information-seeking creatures by nature is further supported by human studies that show that people freely organize their surroundings to maximize information intake, an observation that has led to formalized theories of information foraging.[19] From this perspective, engaging in behaviors that are intended to maximize exposure and consumption of new information, but end up causing interference, may be thought of as optimal. And so, such behaviors may be reinforced despite their negative consequences in other domains of our lives. Since humans seem to exhibit an innate drive to forage for information in much the same way that other animals are driven to forage for food, we need to consider how this "hunger" is now fed to an extreme degree by modern technological advances that deliver highly accessible information; yet another reason for why we are ancient brains living in a high-tech world.

Insights from behavioral ecology, a field that explores the evolutionary basis of behavior by studying interactions between animals and their environments, shed further light on our interference-inducing behavior. An important contribution of this field has been the development of *optimal foraging theories*. These theories are built on findings that animals do not forage for food randomly, but rather optimize their foraging activities based on a powerful drive to survive. Shaped by natural selection, foraging behaviors that successfully maximize energy intake are selected for and thus persist over time. Foraging theory has resulted in mathematical models that can be used to predict the actions of animals given their environmental conditions; that is, they describe how an "optimal forager" would behave in any given situation. Although real-world behaviors certainly deviate from predictions made by these models, these models are frequently not far off the mark and have served as useful tools to understand the complex interplay between behavior and environment. And so, if our interference-inducing behaviors can be thought of as optimal from an information foraging perspective, then optimal foraging theory may help explain the Distracted Mind.

In 1976, evolutionary biologist Eric Charnov developed an optimal foraging theory known as the "marginal value theorem" (MVT), which was formulated to predict the behavior of animals that forage for food in "patchy" environments.[20] Patchy environments are those in which food is found in limited quantity in discrete clumps or patches, with resource-free areas in between. This type of environment, frequently occurring in nature, requires an animal to travel from patch to patch when food resources within a patch become depleted over time. Think of a squirrel foraging acorns in a tree. As the squirrel continues to consume acorns, their availability diminishes, resulting in fewer nuts to eat. At some point, the squirrel is better off taking the time and energy to travel to a new tree to find more food rather than to continue foraging in the increasingly barren tree. MVT models predict how much time an animal will spend in a current patch before moving to a new patch, given environmental conditions.

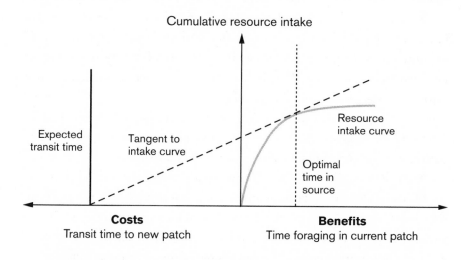

Figure 1.2
A graphical representation of the marginal value theorem, an optimal foraging model that describes the cost–benefit relationship of an animal foraging in a patchy environment.

Without diving into the mathematical details that underlie MVT, we can still understand and even apply the theory by examining a graphical representation of the model. The accompanying figure depicts a cost–benefit relationship on the x-axis, with the benefits accruing with increased "time foraging in current patch" (increasing to the right) and the costs accruing with increased "transit time to new patch" (increasing to the left). An animal, driven by an innate instinct to survive, attempts to maximize its "cumulative resource intake" when foraging (increased up the y-axis). The key factor in the model is designated in the figure as the "resource intake curve." It reflects the diminishing returns of foraging in the same patch over time (represented by the curved line). The cumulative resource intake does not increase linearly or eternally as time foraging in the current patch increases (i.e., the nuts run out). If an animal has knowledge of the factors that underlie the shape of the "resource intake curve" (i.e., an impression of the diminishing benefits of remaining in a patch as they continue to feed), and also knowledge about the "expected transit time" to get to a new patch, then the "optimal time in source" can be calculated as the intersection between a tangent line connecting the "expected transit time" and the "resource intake curve," as illustrated in the figure. And so, if our squirrel is inherently aware that the available acorns are diminishing in his tree and that there is another tree just across the meadow that usually has a lot of acorns and won't take too much time to reach, he would move from the current tree to the new one. This model has now been validated in several animal species, such as the foraging behavior of the great tit and the screaming hairy armadillo.[21]

Now, let's consider the MVT and replace foraging for food resources with foraging for information resources, and insert *you* as the information-foraging animal. Here, the patches are sources of information, such as a website, an email program, or your iPhone. Note that each of these patches exhibits diminishing returns of resources over time as you gradually deplete the information obtainable from them, and/or you become bored or anxious with foraging the same source of information. And so, given both your inherent knowledge of the diminishing resources in the current patch and your awareness of the transit time to reach a new information patch, you will inevitably decide to make a switch to a new information patch after some time has passed. Thus, the model reveals factors that influence our decisions about how long we fish in a particular information pond before moving on to fish in the next pond. The MVT can be successfully applied to human information foraging, and the optimal time of staying versus switching in an information patch may even be calculated mathematically and validated in laboratory and field studies. Although beyond the scope of this book, it will be interesting to see how other scientists empirically address this hypothesis.

Optimal foraging theories have already been applied to human information foraging to help us understand how we search the Internet and our own memories, as well as how scholars and physicians search for information.[22] To our knowledge, such theories have not been used to address the critical question of why we engage in interference-inducing behaviors, even when they are self-destructive. In chapter 9, we apply the MVT model to explore the factors of our high-tech world that influence our information-foraging behavior. We will show that because of unique aspects of modern technology, the manner in which many of us behave may not be considered optimal any longer, even from an information-foraging perspective. In chapter 11, we take this discussion a step further and use the model to formulate a plan of how we can modify our behavior to minimize the negative impact of technology on the Distracted Mind, and thus improve the quality of our lives. We will present strategies in such a way as to avoid missing out on all of the benefits of modern technology. But first, let's take a deeper dive into the underlying basis of our Distracted Mind to stimulate a more informed discussion of what has occurred with the introduction of modern information technology.

2 GOALS AND COGNITIVE CONTROL

THERE ARE TWO equally valid perspectives by which to conceptualize that magnificent organ tucked between our ears: as the brain—the most extraordinary information-processing system and complex structure in the known universe—and as the mind—the emergent higher-order function of that biological machine, the very core of our identity and consciousness. The brain's integration of lightning-fast parallel processing with immense storage prowess is truly astounding, from the identification of complex stimuli within tenths of a second, to the association of events separated by decades, to the storage of a billion bits of data over the course of its operational life—more than 50,000 times the information housed in the Library of Congress.[1] Structurally, the brain is in a class of its own, with over one hundred billion processing units (neurons)—on the order of stars in the core of the Milky Way galaxy—intricately interwoven by hundreds of trillions of connections (synapses) into a distributed network of truly staggering complexity. But perhaps the most impressive feat of the human brain is its functional offspring: the human mind. Despite centuries of academic thought and research on this topic, we still find the most effective way to conceptualize the wonder of the mind is to fully appreciate that it is the essence of every emotion you feel, every thought you have, every sensation you experience, every decision you make, every move you take, every word you utter, every memory you store and recall ... *in the truest sense, it is who you are.*

And yet, despite all this, the human mind has fundamental limitations when it comes to our ability to use cognitive control to accomplish our goals. This makes us vulnerable to goal interference, which in turn has negative consequences for many aspects of our lives. Let us explore the inner workings of our mind to understand why we are so susceptible to goal interference—the underlying cause of the Distracted Mind.

THE PERCEPTION-ACTION CYCLE

Let's begin by turning back the clock to peer into our evolutionary past and ponder how goals themselves emerged as a function of the human brain in the first place. If we could observe our very primitive ancestors, we would find that the role of the earliest version of the brain was not mysterious at all. It was there to support the most basic aspects of survival at the level of the individual and the species. Its function was essentially to guide these creatures toward food and mates, and to direct them away from threats. Even if we look back earlier, before the brain existed, to single-cell organisms without a nervous system, we find that the precursor machinery was performing basically the same function. These primordial lives were dominated by an uninspired sequence of events: detectors on their surface assessed chemical gradients of nutrients and toxins in the surroundings that guided the direction of their locomotion. This was essentially a simple feedback loop that transformed sensation into movement. With the evolution of a distributed nervous system, multicellular organisms evolved more complex and dynamic interactions with their surroundings, but the basic function remained at the core: to sense positive and negative factors in the environment and to use that information to guide actions.[2]

Random mutations that altered the brain in such a way as to enhance the effectiveness of this feedback loop won the Darwinian lottery. Fine-tuning this system, whose function was to increase the chances of eating, procreating, and avoiding being killed, was as good as it got when it came to survival of the fittest. The primitive brain and this feedback loop were thus optimized under the steady influences of natural selection. This interaction between brain and environment continued to evolve until it became a "perception-action cycle," which remains at the core of behavior for all modern animals.[3]

The perception-action cycle is fed by sensory inputs from the environment—sights, sounds, smells, and tactile sensations—whose signals enter the brain via an expansive web of specialized nerves. This sensory information is then represented by patterns of neural activity at the back half of the brain on its thin surface known as the cortex. These patterns are sculpted by processes of divergence, convergence, amplification, and suppression, which form complex representations of the external world, or *perceptions*. Across the brain, in the front half, *actions* are generated internally and represented by activity patterns on the cortex. Brain areas specialized for perception and action communicate dynamically with one another via bidirectional bridges that are the building blocks of neural networks. These connections are defined both by the brain regions they connect, as well as how they interact with one another, known as functional connectivity. This is similar to highways being defined by both by the cities they connect and the traffic patterns that join them. Supported by rapid communication across these pathways that stretch between the front half and the back half of the brain, the perception-action cycle spins away endlessly: environmental stimuli lead to perception, which drives an action, resulting in changes in the environment that spawn new perceptions, followed by responsive actions—and on and on the cycle spins.

The perception-action cycles of primitive brains were essentially automatic, reflexive loops. These brains were not so very different in function from that of the precursors of the nervous system in single-celled organisms. Scientists observe that simple organisms, such as worms, detect a relevant chemical trace in the environment and make a beeline toward or away from its source depending on its composition: a precursor of the perception-to-action loop that might be thought of as a sensation-to-movement cycle. Studying the brains of experimental animals has allowed us to dissect the details of the circuits that underlie this cycle and in turn the foundations of the inner workings of the human brain and mind. But one major difference is that primitive brains do not engage in true decision-making processes. That is, there are no higher-order evaluation processes, no goal-setting or goal-enactment abilities that drive their behaviors. They are driven entirely by this reflex: environmental triggers activate sensory neurons via specialized receptors that send signals to appropriate motor neurons to result in prede-termined responses.

Interestingly, we can see perception-action reflexes still at work in all modern animals, as well as in ourselves. The patella reflex, also known as the knee jerk, is a classic example of such an ancient reflex: sensory informa-tion from a tap to your patella tendon travels directly to your spinal cord, and then via a relay system, a motor response is triggered and expressed as an abrupt leg movement. This basic reflex pathway serves a critical role in allowing us to walk without continuous attentional control.[4] You can find other examples of similar reflexes throughout our bodies, such as the pupil-lary reflex, in which your pupils automatically change size in response to variations in light levels, and of course the pain reflex, which results in a rapid withdrawal response to the sharp prick of a needle.

TOP DOWN AND BOTTOM UP

Although these reflexes remain critical to our functioning and survival, the perception-action cycle has undergone major evolutionary modifications. First, both perceptions and actions have become much more complex. Human perceptions have advanced beyond simple sensations to involve multifaceted interpretations of sensory stimuli. They are also integrated with contributions from memories of past events, such that they now embody the context from which they were previously experienced. Actions are no lon-ger limited to simple motor responses, but involve higher-level output and nuanced expressions that might not even be recognized as "actions," such as language, music, and art.

But an even more profound evolutionary modification to the percep-tion-action cycle was the evolution of a mechanism that "broke" the cycle, so that the relationship between input and output were no longer always automatic and reflexive. Although perception-action reflexes continue to offer us critical survival benefits, and are thus preserved at many levels of our nervous system, they now exert a diminished influence on our most complex behaviors. Indeed, it is this break in the perception-action cycle that created the golden opportunity for the evolution of goals, which are perhaps the defining unique characteristic of the human mind.

This most remarkable milestone in our brain's evolution, the insertion of a break in the perception-action cycle, can be conceptualized more appropriately as a time delay between perception and action—a *pause*. During this pause, highly evolved neural processes that underlie our goal-setting abilities come into play—the *executive functions*. These abilities of evaluation, decision making, organization, and planning disrupt the automaticity of the cycle and influence both perceptions and actions via associations, reflections, expectations, and emotional weighting. This synthesis is the true pinnacle of the human mind—the creation of high-level *goals*.[5]

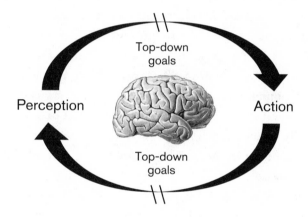

Figure 2.1

A diagram of the perception-action cycle, depicting top-down goals in humans as disrupting reflexive responses to our environment. The hash marks indicate pauses in the perception-action cycle.

Goals are internally generated plans that guide our actions, allowing us to choose how we respond to incoming perceptions based on *evaluations* that we engage in and *decisions* that we reach. This results in many of our actions no longer being automatic—or at least not completely reflexive. Of course many of our actions still are. If a child pinches your arm, you will invariably recoil, fulfilling the pain-withdrawal reflex. But you are then unlikely to strike back. You are able to pause, engage in goal-setting processes, *evaluate* the act as nonmalicious and the young perpetrator as a nonthreat, to reach a *decision* that a violent response is unwarranted and inappropriate. This allows you to suppress retaliation, which in a less evolved creature might have been a reflexive counterstrike out of self-defense. As we will discuss later, the pause is not only one of the more recent evolutions of the human brain, but also the latest to develop in each of our lives. Children with underdeveloped goal-setting abilities often do strike back in this situation, much to the dismay of parents around the world.

Most of us are inherently aware that our goals influence our actions, as we've described, but it is less obvious that our goals also shape the way in which we perceive the world. Neuroscience research has helped us understand that perception is not a passive process; sights, sounds, and smells of the world do not simply flood into our brain. Rather, the inward flow of information is sculpted and pruned by goals in much the same way that our actions are, resulting in our perceptions being an interpretation of reality and not a veridical representation. Those flowers you decide to pay attention to actually do look much redder to you and smell much sweeter than the ones you chose to ignore. Goals thus influence both sides of the cycle, perception and action.

However, just because we have evolved goal-setting abilities does not mean that these are now the only influence on the perception-action cycle. These internal, goal-directed, *top-down* influences occur at the same time as external, stimulus-driven, *bottom-up* influences to modulate our perceptions and actions. The bottom-up forces are the same as they have always have been: novelty and saliency. Stimuli that are unexpected, dramatic, and sudden, such as a flash of light or a loud crash, or important, either innately or based on memories of previous experiences, such as your name, command bottom-up dominance in our minds, independent of our top-down goals. These bottom-up influences are essentially those same driving forces of the perception-action cycle that converted perception into action to ensure the survival of our ancestors, and thus are yet another retained aspect of our ancient brains. They live on as a major influence of our perception-action cycle, and as we will see, play a major role in the story of the Distracted Mind.

Note also that humans are not the only species with top-down, goal-setting abilities. Other animals have evolved the ability to formulate and enact goals. Some even do so in a sophisticated manner; for example, great apes and corvids, such as crows and jays, exhibit the ability to create simple tools to accomplish future tasks.[6] But even this impressive behavior is nowhere near the same order of complex, interwoven, time-delayed, and shared goal-setting activities that humans routinely exhibit. Human goal-setting feats have allowed us to become proficient at exerting tremendous control over how we interact with our world. This in turn has led to wondrous human creations, such as complex language, society, and technology. The freedom of not being slaves to bottom-up influences has released us and created the unique opportunity for the generation and fulfillment of creative insights that are the hallmark of innovation. This is in contrast to the mental landscape of many other animals, whose interactions are dominated by reflexive responses to bottom-up influences.

Despite this, animal behaviors are often attributed human-like goals. This is an act of anthropomorphization—the assigning of human qualities to other entities. The interactions between most animals and their environment are fundamentally different from ours in this important regard. That eagle diving out of the sky to swoop up a field mouse in its talons is not being "vicious." Those ants marching in an intricate pattern across your kitchen floor are not being "devious." Their abilities are no less amazing—more so, in many ways—but they are still largely servants to the bottom-up world in which they live. It is their sensitivity to novel and salient stimuli with correspondingly rapid and reflexive responses to the environment that has allowed them to survive in such a demanding and competitive world. This is their asset, and in many ways our lack of bottom-up sensitivity can be viewed as a human deficit.

Consider the dramatic stories that emerged following the 2004 Indian Ocean earthquake off the west coast of Sumatra. This undersea earthquake resulted in a tsunami claiming the lives of approximately 230,000 people in fourteen countries. Interestingly, reports noted that as humans stood transfixed, or even approached the receding tide that preceded the tsunami out of curiosity (top-down goals), other animals took off to higher grounds before the destruction ensued. The owner of the Khao Lak Elephant Trekking Centre in Thailand reported that elephants broke free of their restraints, ignored their trainers' commands, and ran for the hills five minutes before the area they had been standing on was destroyed. As Bill Karesh of the Wildlife Conservation Society commented: "We know they have better sense of hearing; they have better sense of sounds; they have better sense of sight. And they're more reactive to those signals than we tend to be." They also respond to salient signals detected in the behavior of other animals more readily: "If they see birds flying away, or if they see other animals running, they're going to get nervous too." NBC News reported that "when the tsunami struck in Khao Lak, more than 3,000 human beings lost their lives. But no one involved with the care of animals can report the death of a single one." Goson Sipasad, the manager of the Khao Lak National Park, reported that "we have not found any dead animals along this part of the coast." Amazingly, four Japanese tourists survived by being carried to the safety of the surrounding hills by elephants they happened to be riding at the time.[7]

As mentioned, we do retain bottom-up sensitivity to the environment, as it continues to offer a survival advantage, even in our modern world. Without the detection of any bottom-up signals, we may not notice the smell of unexpected smoke or the honk of a car horn as we step out into the street. How then does this preserved, ancient system interact with our top-down goals? This topic is one of great complexity, and consumes the research efforts of many laboratories, including the Gazzaley Lab; limitations in our ability to balance these two influences are a major factor for the Distracted Mind. We will explore this interaction later in the book. For now, it is useful to consider how this imprecise integration of bottom-up influences and top-down goals leads to a fascinating dance of goal interference that occurs in our daily lives. This dance is rich in both distractions and interruptions, both externally and internally generated, as illustrated in this everyday scenario:

You are driving along the highway, focusing on a complicated traffic situation unfolding before you, when you can't help but notice a text message has arrived (bottom-up influence). The vibration against your leg jolts you from the top-down goal of searching for your upcoming exit, and, of course, driving safely in traffic.

This external distraction is a clear example of goal interference. If you are not effective in immediately suppressing this bottom-up influence from entering your consciousness, it will inevitably lead to evaluation and decision-making processes. Even this minor act will have converted an external distraction into an internal interruption. And should you then decide to actually check the text message, you would truly derail your goal of driving safely without interference. It will result in an even more disruptive (and dangerous) external interruption: the infamous texting and driving, which compounds the issue by removing your eyes from the road.

And so you decide to ignore the text and remain focused on your primary top-down goal of driving safely. But the vibration lingers in your mind, and it begins to feel like a burning ember in your pocket that is accompanied by rising anxiety—who is texting at this hour, and what do they have to tell you? You attempt to push it from your consciousness, but it persists.

Now you have a new source of interference, an internal distraction that involuntarily threatens to dislodge you from your driving goals.

You finally give in and make the decision to divert some attention away from driving to think about who might be trying to reach you.

Now you have another internal interruption.

… and the next thing you know, you completely missed your exit, because although your eyes remained firmly on the road, your brain did not. This blunder requires you to consult your smartphone to figure out how to get back on track, thus providing yet another source of goal interference: external interruption.

And there you have it, the interference dance in full effect: external distraction → internal interruption → internal distraction → internal interruption → external interruption. On and on it goes, as the interplay of bottom-up influences and top-down goals leads to constant distraction and interruption while we are trying to accomplish our goals. Is it any wonder that 80 percent of car accidents and 16 percent of highway deaths are the consequences of distracted driving?[8] *We are all cruising along on a superhighway of interference.*

And yet, we do have the ability to make decisions. Our actions are not a straightforward balancing act between the salience of an already initiated action (driving) and the novelty of a new input from the outside world (text message), as is true for many other animals. Consider the following example in the animal world:

A fox approaches a stream for a drink of water and catches the scent of a predator in the air. The fox immediately turns and sprints rapidly toward the forest cover.

This action is largely a straightforward weighting of a new stimulus-response plan against an existing one, which in this case resulted in a reflexive retreat response. Although the fox is not making a true, goal-directed decision, it is one that is critically valuable for its survival. Goal interference by bottom-up influences is different for us than other animals; our responses are rarely the product of such a straightforward weighting between the saliency of two stimuli. More often than not, it involves a pause, as evaluation and decision-making processes churn away and lead us to the creation of goals that permit us to intentionally suppress salient information. Top-down goals are so strong that they may even lead us to be completely unaware of powerful bottom-up stimuli that are more salient than an ongoing activity, as shown in this example:

You are driving along and miss your exit because you are thinking about that ignored text message; or you walk right past a good friend on the street as you stroll along talking on your phone.

It is clear that goal-directed behavior is a complicated act that in many ways defines what is most human about us. We have this unique ability to generate high-level, time-delayed goals that allow us to sculpt our interactions with our environment in a powerful and personal way. This brings us to the next question: How do we go about enacting our top-down goals once we have established them?

CONTROL PROCESSES

Everyone knows what attention is. It is the taking possession by the mind, in clear and vivid form, of one out of what seem several simultaneously possible objects or trains of thought. Focalization, concentration, of consciousness are of its essence. It implies withdrawal from some things in order to deal effectively with others, and is a condition which has a real opposite in the confused, dazed, scatterbrained state which in French is called *distraction*, and *Zerstreutheit* in German.

—WILLIAM JAMES [9]

As described, brain evolution led to the insertion of a critical time delay that disrupts the reflexive nature of the perception-action cycle so that the neural processes that underlie goal setting—evaluation and decision making—may be engaged. This extraordinarily important pause in the cycle disrupts our reflexive responses to environmental stimuli, allowing us to generate top-down goals. These goals exert influence over both our perceptions and actions, which compete with those powerful bottom-up forces. But setting goals is not enough to affect our lives and the world around us; we need to enact our goals. The mediators of our top-down goals comprise another amazing collection of abilities that fall under the umbrella of cognitive control. This includes three major faculties: (1) attention, (2) working memory, and (3) goal management, each consisting of subcomponent processes. It is this battery of cognitive control abilities that allows us to interact in our complex world in a dynamic and goal-directed manner. And with varying degrees of success, it is what allows us to resist the negative impact of goal interference. To understand the essence of the Distracted Mind, we need to carefully dissect these core abilities of our mind and appreciate their strengths and limitations.

ATTENTION: THE SPOTLIGHT

"Attention" is likely the most widely used term in cognitive science. The general public and practitioners from diverse fields of education, philosophy, mental health, marketing, design, politics and human factors, wield this word freely. This is despite the fact that most users do not have a deep understanding of the construct of attention. This is likely because it is so integral to our daily lives that the role it plays in our real-world functioning seems intuitive. Even William James, the father of American psychology, famously remarked in his landmark 1890 textbook that "everyone knows what attention is." But do not allow this pervasive use of the word in our vernacular to fool you into thinking that attention is a straightforward concept. On the contrary, it is a complex set of integrated processes with multiple subcomponents, and it has remained a dominant focus of scientific investigation in psychology and neuroscience over the last century.[10]

To unravel the depth and breadth of what attention is, the first step is to appreciate its most fundamental feature: *selectivity*. This is what allows us to direct our brainpower—our neural resources—in a focused manner. Like an archer shooting a cognitive arrow at its goal, selectivity maximizes the effectiveness of our neural processing and in turn optimizes our performance. It is a precision actuator that allows us to selectively bias future perceptions and actions in alignment with our goals. Let's turn back the clock again and consider a scene from our human past to appreciate selective attention during its formative period:

One of our thirsty ancestors is prowling through a deep forest and emerges into an unfamiliar clearing where he spots an enticing stream. Success! But, despite the powerful, actually reflexive, drive to drink, he does not charge forward. Rather, he pauses ... suppresses the impulse, evaluates, reaches a decision, and formulates a goal. Previous experience in this forest has taught him that where there is water, there is often also a highly effective predator hiding nearby—the jaguar. This is where that critical pause in the perception-action reflex of "see water → drink water" allows him to undertake an initial rapid evaluation of the situation to determine if this setting deserves more careful assessment. This leads him to decide that there may indeed be an invisible threat. He then advances to establishing the top-down goal of carefully evaluating the safety of approaching the stream.

To accomplish his goal, he now engages his selective attention to focus his hearing on detecting a specific sound that he knows a jaguar makes when lying in wait for prey: a deep, nearly inaudible, guttural grumbling. He selectively directs his vision toward detecting the characteristic pattern and color of the jaguar: a stippled orange and black.

And knowing the distinctive odor of this creature, he selectively focuses his olfactory sense on its characteristic musky smell. In addition, he has the insight to know that this particular predator tends to hunt in the thick brush that lies along the left bank of the stream, and so he directs this multisensory selective attention like an arrow at that specific location in space; he fires and waits for a signal.

This scenario demonstrates how attentional selectivity can be directed at many different aspects that constitute the environment. Attention was directed at sensory features, such as sounds and smells (feature-based attention), a specific location in space (spatial attention), and also the jaguar's overall form (object-based attention). In addition, he could have chosen to deploy his attention to a selective point in time (temporal attention):

Our ancestor knows that the predator may be startled into moving and reveal himself by an unexpected event, like a splash in the water nearby. And so, as he throws a rock into the stream, he directs his attention to a select moment in time—right when the stone hits the water. And there it was: subtle movement detected in the brush; time to get out of here.

This allocation of selective attention resources converging across multiple domains, all focused on a singular objective, increases his likelihood of accomplishing the goal of detecting a hidden jaguar.

Let's review the sequence of events as they played out: An action takes place (our ancestor steps into a clearing area) → a perception follows (he sees a stream) → a pause takes place to derail a perception-action reflex (he does not approach the stream to take a much-needed drink) → goal-setting processes are engaged (he evaluates the scenario to decide what to do) → a goal is formed (he will more carefully assess the danger of this situation) → cognitive control abilities to enact the goal are engaged (he brings selective attention processes online to detect a hidden jaguar) → a new perception takes place (he detects a slight rustling movement on the left bank of the stream in response to a thrown rock) → a new action occurs (he retreats to the forest).

You can see clearly in this scenario how selective attention is a powerful tool by which our goals assert influence on the perception-action cycle. In this case, it influenced our ancestor's perception to facilitate a hasty retreat, but in another scenario it might also have led to suppressing actions, such as inhibiting a retreat to the forest in response to a bird abruptly emerging from the brush. Suppression of action, known as "response inhibition," is another critical aspect of the attentional selectivity that is directed at the action side of the cycle. In fact, the initial pause that occurred after the stream was spotted is an example of response inhibition.

Selectivity can be thought of as the "spotlight" in our cognitive control toolkit. It enables fine-tuning of processing across all sensory domains (audition, vision, olfaction) for relevant features (a low grumbling sound, orange stripes, a musky smell), relevant locations (brush on the left bank), and relevant moments in time (when his rock hit the water). If a musky jaguar smell enters our ancestor's nostrils, the smell's processing is exaggerated and becomes more noticeable than if selective attention were not engaged. As described, selective attention is a spotlight not just for perception, but also for action: it selectivity fine-tunes our responses based on our goals. This is notable for the suppression of responses that might otherwise be reflexive. In a similar way, selective attention also involves suppression of perceptions that are in the dark, outside of the spotlight, also known as the act of ignoring:

While searching for telltale signs of a jaguar, our ancestor ignores the sounds of rodents scurrying along the forest floor and birds rustling leaves in the trees.

Perceptual inhibition is important to decrease goal interference in our ancestor's attempt to detect the jaguar's low, grumbling snarl. Attending and ignoring can be thought of as the lens by which the selectivity spotlight is focused to achieve its precision that allows our thirsty ancestor to effectively interpret and then act on these subtle cues in his surroundings. In this case, a hasty retreat into the forest after he detected a jaguar worked out a lot better than if he had made the reflexive approach to the stream to get a drink. Selectivity, sharpened by the dual processes of focusing and ignoring, is a critical element by which attention enables our top-down goals.

Selective attention allowed our ancestor to accomplish his goal of detecting the jaguar and responding optimally. You can see how it increased his chances of surviving, thus illustrating the natural pressures that promoted the evolution of this ability in the first place. However, the effectiveness of attention relies on much more than whether the spotlight is focused or not. It is also critical *when, where,* and *how long* the spotlight is wielded. These three aspects of attention that build on its selectivity are known as *expectation, directionality,* and *sustainability.*

When we use our spotlight.   As illustrated, we engage selective attention when we perceive a stimulus as well as when we act upon that information. But, critically, we can also employ selective attention before a stimulus is even present or before an action occurs. In other words, we can set the selectivity machinery in motion prior to perception or action, in anticipation of it. This is often referred to as *expectation.* It is a key factor in how and when we use our selectivity spotlight. It is what allows us to transition from the internal world of our goals to the external world of our perceptions and actions. When our ancestor first paused, there was no sight, sound, or smell of a jaguar prompting him to do so. What he did was generate an expectation of future events based on his previous experiences of what sights, sounds, and smells might occur. Expectation is a critical factor in optimizing our performance by enabling knowledge of past events to shape our future. In many ways, our brains live in the future, using predictive information to bias both incoming stimuli and outgoing responses.

Where we use our spotlight.   Directionality is another important feature of selective attention. We can direct our limited cognitive resources to stimuli in the environment, as described in the above scenario—a sound, a place, a color, a smell—but we can also aim it internally at our thoughts and emotions. Just as is true for external selective attention, our ability to control internal attention allows us to attend to relevant or ignore irrelevant information in our minds based on our goals. The scope of internal-directed attention is just as broad as it is for the external world; we can direct our attention toward searching memories and/or focusing on feedback from the body, such as a pain in your side or the rumble of your hungry stomach. Likewise, it is often important to selectively ignore internal information, such as suppressing sadness at a time when you need to remain upbeat, or suppressing a recurrent thought that is interfering with your current activities—or in our ancestor's case, suppressing a strong desire for a drink. Failure to suppress internal signals that are irrelevant to your current goals is a major source of internal distraction.

How long we use our spotlight.   Another critical factor when using selective attention is our ability to sustain it. This is especially true in situations that are not engaging, or even boring. You can imagine how challenging it might be for our ancestor to sustain his selective attention as time goes on if there were no sign of a jaguar. Perhaps that may have been appropriate, but maybe based on previous experiences of how patient jaguars can be, it may be premature for him to move quickly and critical for him to hold his attention for longer periods of time. In this case, the need to sustain attention, even beyond comfort levels, could save his life. Air traffic controllers experience a modern-day scenario of this on a daily basis with similar high-stake consequences of a failure to sustain attention. They routinely need to sustain their attention for hours to detect meaningful irregularities in a boring, repetitive pattern; they must be vigilant. An even more common example is a teenager trying to sustain selective focus on an algebra lesson during class. As with all aspects of cognitive control, building these abilities is an important aspect of brain development.

Attention, complete with selectivity, expectation, sustainability, and directionality, is an undeniably powerful mediator of our goals. But it is only one of the tools in our cognitive control toolkit.

WORKING MEMORY: THE BRIDGE

Living at the junction of our external and internal world is another critical component of cognitive control—*working memory*. This ability allows us to hold information actively in mind for brief periods of time to guide our subsequent actions.[11] Our working memory kicks in when a stimulus is no longer present in our surroundings. In many ways, working memory is the quintessential instrument of the pause: it is the *bridge* between perception and future action. If we insert a time delay between what we perceive and how we respond to it, then there needs to be a connecting mechanism to bind information in time across the delay. This is where working memory comes in. Similarly to how expectations connect our internal goals to our external world, working memory bridges the leap from perception to action. It allows us to flow smoothly through events in our lives that are separated by brief time delays, all the while maintaining a sense of continuity. Some view this as a type of attention that is directed internally rather than externally, which is a reasonable point of view.[12] Putting semantics aside, working memory is a critical cognitive control tool that is necessary for goal enactment. Let's consider its role in our scenario:

Our ancestor is walking through the forest when he steps into a clearing and spots the stream. He immediately darts behind a large tree for cover, so that he can search for a potential predator from a safe position. In the time that passes between when he sees the stream until he peeks his head out from behind the tree, he does not forget the position of the stream or the thick brush lying along the left bank. In fact, he sees it all clearly in his mind's eye as he is ducking for cover, almost as if the scene were still in his view. And so, when he pokes his head out, he immediately focuses his selective attention at exactly the right spot along the bank.

Working memory bridged our ancestor's perception of the scene in front of him with his subsequent action of searching for the jaguar. It allowed him to maintain a representation in his mind of all the sights, smells, and sounds around him. If something changed in his sensory landscape after he looked away for a moment, or after he threw a rock into the stream, he could compare the new scene to the previous scene that is no longer present but is actively held in his working memory. This illustrates the role that working memory plays in enacting our goals, as well as its close interplay with selective attention. Working memory forms connections throughout our day that are transient and usually take place without our conscious awareness, so it may be challenging to appreciate what your life would be like without this ability. But make no mistake: working memory is a critical aspect of cognitive control that is essential to everyday functioning. If it did not exist, we would forever be disconnected, floating in a disjointed way from one event to the next. Just imagine a conversation between you and your friends without working memory … it would not go well.

Working memory is often used interchangeably with short-term memory. This is in contrast with long-term memory, which involves another process known as consolidation that yields enduring memories lasting from minutes to many years. Working memory is often described as the brief retention of information in our mind's eye, but that is not the full picture because it might just as well be our mind's ear or our mind's nose. We have the ability to hold all types of information in mind, including verbal information, abstract concepts, thoughts, ideas, and emotions that are not even sensory in nature. Furthermore, this brief retention of information is only one aspect of working memory. Many conceptualizations of working memory over the years have brought us to a consensus that working memory involves both processes that maintain representations of information in mind and processes used to manipulate this information. Working memory is not a passive process, but an extremely active one. For example, our ancestor may not just be holding the scene in mind as he ducks behind a tree, but also judging the distance to the stream and the thickness of the brush, as well as comparing it to previous streams he has encountered. All of this aids him in reaching his goal successfully.

GOAL MANAGEMENT: THE TRAFFIC CONTROLLER

Our brains might very well have evolved so that we tended to generate only one goal at a time, such as is true for most animals, even other primates.[13] But that is not what happened. Humans frequently decide to engage in more than one goal-directed activity at a time, or to switch back and forth between multiple goals. The common terms used for these behaviors are "multitasking" and "task switching," respectively, both examples of *goal management*, and another core aspect of cognitive control that serves as a meditator of our top-down goals. Goal management comprises a set of abilities that enables us to tackle more than one goal over a limited time period. As our actions became more than merely reflexive responses to bottom-up stimuli, the need invariably arose for us to be able to manage multiple goals in overlapping timeframes. If the outcome of an evaluation and decision-making process leads to the generation of competing goals, what happens then? What happens if during the enactment of a goal, a new goal is formulated? Goal management serves the critical function of our mental *traffic controller*. Of course, success in actually accomplishing multiple goals is dependent upon integrating goal management with attention—the spotlight—and working memory—the bridge—in a fluid and flexible manner. Let us return once again to our ancestor to see how goal management combines with our other cognitive control abilities:

While on his forest trek, our ancestor emerges into a clearing and discovers the stream. At this point in his perception-action cycle, the innate reflexive response to rapidly approach the water is disrupted by a pause and a quick dart to the cover of nearby trees. Evaluation and decision-making processes result in his formulating a decisive goal to search for a hidden jaguar in the brush on the bank of the stream. He proceeds to engage all his cognitive control abilities to enact this goal. Attention is employed like a spotlight to surgically dissect the scene before him and search for relevant sensory features, while his working memory acts as a bridge between his initial perceptions of the scene and his new perceptions, as well as his subsequent actions. His initial surveillance yields no signs of danger, but our clever ancestor remains suspicious. His long-term memories contribute to his survival. He is aware that at this time of year and at this elevation, chances of encountering jaguars are high, and so he establishes a new goal to throw a rock into the water near the brush and try to induce movement. This new goal is clearly in support of the overarching objective of jaguar detection, but involves a whole new cascade of perceptions and actions, such as finding an appropriately sized stone and making an accurate throw, which need to be managed in parallel with his ongoing goals of remaining sensitive to jaguar signs. And so, after he ducks behind the tree for cover he enacts the new goal of searching for a rock. While this is going on, he maintains in working memory his original goal of detecting jaguar-specific smells, sights, and sounds, and also the details of the scene he witnessed when first arriving at the site. All of these processes must be managed simultaneously. Goal management allows him to navigate this complex and dynamic situation, which eventually leads him to the conclusion that there is a jaguar hidden in the brush and his need for a drink will just have to wait.

Goal management becomes especially critical when we need to engage in more than one activity at a time that draws on the same limited resource pool. This is clearly true for many perception and action processes, but also for cognitive control processes. For example, our ancestor's search for a rock involves his paying selective attention toward the forest floor to locate a stone large enough to make a splash but not so large that it won't reach the stream. This search for the rock competes directly with his ongoing selective attention of searching for the color and pattern of the jaguar's coat. Because these two activities both involve vision, it is obvious that that search for a rock on the forest floor will blunt his effectiveness in spotting a hidden jaguar in the brush, even if it happened to move during this time. In this case, goal management involved switching between visual selective attentions goals, even though he was not aware that he was doing it. This is not dissimilar to the texting and driving scenario in that it involves direct competition for vision. But it is important to realize that the competition between two selective attention goals does not only occur when there is competition for the same sensory resources. His ability to hear and smell the jaguar is also diminished when his selective attention is directed toward a search for a rock. This is because two tasks that demand cognitive control, even if they are not competing for the same sensory resources, require mental task switching. In the next chapter, we will explore brain mechanisms that underlie all aspects of cognitive control, and then in chapter 4 we will discuss the limitations in our cognitive control abilities and how these limitations yield the Distracted Mind.

3 THE BRAIN AND CONTROL

GIVEN THE FUNDAMENTAL role that cognitive control plays in our daily lives, it is understandable that one of the most active areas of neuroscience research has been the study of how our brains enable these core abilities. Over the last several decades, our success in studying the human brain has been greatly expanded, as an extensive array of powerful tools have been invented that allow neuroscientists to noninvasively scrutinize its structure, chemistry, and function in controlled laboratory environments. Prior to the development of these technologies, most of what we learned about human brains was extrapolated from studies of the brains of other animals and the findings of psychological research. While these approaches certainly have led to fundamental advances in understanding our brain and continue to inform us, key gaps remain in deciphering the basis of unique aspects of the human brain. The acronyms of the technologies that work across different resolutions in space and time read like alphabet soup: PET (positron emission tomography), MRI (magnetic resonance imaging), EEG (electroencephalography), TMS (transcranial magnetic stimulation), TES (transcranial electrical stimulation), MEG (magnetoencephalography), and NIRS (near infrared spectroscopy). These technologies offer us diverse and powerful approaches for human neuroscience to build on foundations from experimental animal and human psychology research. They enable us to address the critical question of how our brain's anatomy, chemistry, and physiology lead to the emergence of our minds. This pursuit has been tremendously successful in fostering a rich understanding of the neural underpinnings of cognitive control, as well increasing awareness of their significant limitations.

We've learned two important lessons along the way. First, one region in our brains evolved the most from that of our ancestors to become a major mediator of cognitive control. This brain region is known as the *prefrontal cortex*. Second, although the prefrontal cortex clearly plays a critical role in cognitive control, it does not function in isolation. Rather, it acts as a node in a vast interconnected web of multiple brain regions, known as a *neural network*. So, while it may be true that the prefrontal cortex enables cognitive control, the underlying basis of how cognitive control is achieved is via the interactions across an intricate network that connects the prefrontal cortex with many other brain regions. Let's begin by discussing the role of the prefrontal cortex and its networks in cognitive control and then the physiological mechanisms that underlie cognitive control.

THE PREFRONTAL CORTEX

It is now well established that the prefrontal cortex serves as a central hub for both our goal-setting and goal-enactment processes, and thus holds the title as the defining brain structure that makes us most human.[1] However, the function of the prefrontal cortex, which resides in the most forward regions of the frontal lobes, directly behind our foreheads, has historically been shrouded in mystery; only relatively recently has it given up some of its many secrets. Interestingly, not all areas of the frontal lobes have been so resistant to revealing their function. For example, the area farthest back in the frontal lobes, closest to the center of our heads, now known as the motor cortex, was documented clearly in the late 1800s to be responsible for movement.[2] This was revealed by experiments showing that destruction of one side of the motor cortex resulted in paralysis of muscles on the opposite side of the body. Electrical stimulation of these brain regions had the opposite effect and caused movement on the other side of the body. Researchers soon learned with further experimentation that these effects were the result of influencing neurons of the motor cortex that projected to neurons on the other side of the spinal cord, which in turn stimulated muscles on that half of the body. As investigations advanced forward along the frontal lobes, they reached brain regions that were also involved in movement, but at a higher level of motor planning. These areas are now known as the *premotor cortex.*

As research on the frontal lobes continued to advance forward toward the forehead, things became murkier. Studies that attempted to uncover the function of the prefrontal cortex using the lesion and stimulation approaches that were so successful for understanding the motor areas yielded unclear results. While researchers discovered the function of some prefrontal sub-regions, such as the lower left area of the prefrontal cortex being critical for language expression, large expanses seemed to have no clear function at all, leading some scientists to consider these to be the "silent lobes" of the brain.[3] This was at odds with the prefrontal cortex being noteworthy as an expansive and prominent region of the human brain and may have contributed to the urban myth that we use only 10 percent of our brains.[4] Of course, we now appreciate that we use *all* of our brain, even though some of its functions are complex and not immediately obvious. It would be surprising indeed if the most complex structure in the known universe had vacant office space. And yet, the mystery of the function of prefrontal cortex stretched on for decades. Interestingly, a random event that occurred in the mid-nineteenth century opened our eyes to the true nature of its function. This was not the result of a laboratory experiment, but rather the consequences of a traumatic injury to an individual, who likely represents the most important medical case in the history of neurology.

On September 13, 1848, at 4:30 p.m., Phineas Gage, a twenty-five-year-old railroad construction foreman, was working on the Rutland and Burlington Railroad near Cavendish, Vermont.[5] His team was responsible for blasting a path for their new railway. This involved drilling holes in large boulders, adding gun powder, a wick, sand, and finally packing these materials in with a massive iron pipe, known as a tamping iron. Reports suggest that Gage was momentarily distracted while preparing one of the holes (*oh, the irony*) and neglected to add sand. This led to a spark igniting the gunpowder when he struck down with the tamping iron, which in turn caused the three-foot-long, 1.25-inch-diameter, fourteen-pound iron rod to be propelled up through his head, passing below his left eye, and traveling out the top of his skull to land almost one hundred feet behind him. Miraculously, Gage was not killed by this dramatic accident, and seemingly did not even lose consciousness. He is even reported to have remained awake with a gaping hole in his head during his transport in a cart to his boarding house forty-five minutes away, where he received medical care and survived the incident.

As fortunate as Gage was to have not been killed instantly, he did of course experience extensive brain damage from the passage of the tamping iron through his head.[6] Although a careful medical examination of his brain was not performed during his lifetime, which lasted eleven years after the accident, his body was exhumed seven years after his death and his skull subsequently became a source of detailed investigations that used sophisticated imaging methods to determine exactly which brain areas had been damaged. Despite debate that remains over the precise details of his brain injury, what is agreed upon is that regions of his frontal lobes involved in motor control, eye movements, and language were spared by the projectile, but there was widespread destruction of the front and lower regions of his left prefrontal cortex, and perhaps a bit of damage to the right prefrontal cortex.[7] The damage included the fibers that travel under the cortex, called the "white matter," which connects these brain regions with the rest of the brain and form the structural scaffolding for the function of neural networks.

An even more fascinating aspect of this story than Gage's unlikely survival is that the injury did not result in deficits to his basic functioning. He was still able to walk, talk, and eat without difficulty. He was also reported to be "in full possession of his reason." This likely confirmed suspicions of many researchers at that time that the most frontal regions of the prefrontal cortex were not important brain structures. But all this soon changed, as the case of Phineas Gage went on to offer the world a major challenge to this view of the prefrontal cortex. You see, Gage was in fact radically changed by his injury, but in an unexpected manner. Prior to the accident, he "possessed a well-balanced mind, and was looked upon by those who knew him as a shrewd, smart businessman, very energetic and persistent in executing all his plans of operation," as described by his physician John Harlow. In contrast, a shocking transformation occurred to Gage's personality after the injury, as described in an official report twenty years after its occurrence:

He was fitful, irreverent, indulging at times in the grossest profanity (which was not previously his custom), manifesting but little deference for his fellows, impatient of restraint or advice when it conflicts with his desires, at times pertinaciously obstinate, yet capricious and vacillating, devising many plans of future operations, which are no sooner arranged than they are abandoned in turn for others appearing more feasible. A child in his intellectual capacity and manifestations, he has the animal passions of a strong man. ... In this regard his mind was radically changed, so decidedly that his friends and acquaintances said he was "no longer Gage."[8]

Gage was plagued by this new and unsettling personality for the rest of his life, beginning with the loss of his job at the railway, where he was no longer able to perform responsibly, and continuing for a decade of wandering as a vagabond, an existence that seems not to have had much higher purpose. His personality alteration was nothing less than shocking and convinced scholars that the prefrontal cortex played a critical role in emotion, demeanor, and interpersonal skills. Researchers at the time who were interested in the prefrontal cortex found new inspiration in this case and a solid foundation for renewed scientific inquiry into elucidating the role of this brain region as a core mediator of some of the most complex aspects of human behavior.

Interestingly (and most unfortunately), this was not the last medical case to give us a window into the function of the prefrontal cortex. For Gage, the radical personality transformation was the result of a tragic accident, but tens of thousands of others around the world suffered a similar fate as the catastrophic consequences of a medical procedure known as the frontal lobotomy.[9] This procedure involved the purposeful destruction of the prefrontal cortex by physicians as a treatment for a wide range of psychiatric conditions. As you ask yourself the reasonable question of how such an aggressive and often completely unethically applied procedure came to be performed, it may be helpful to realize that this occurred at the intersection of the challenging social and financial times of the early 1900s and our emerging, but still naive, understanding of the prefrontal cortex.

Evidence was accumulating throughout the second half of the nineteenth century that demonstrated the importance of the prefrontal cortex in diverse aspects of human personality. This knowledge was generated at a very difficult time in the history of psychiatry. Since there was essentially nothing that could be done to help the countless number of individuals suffering from severe mental illness, an increasingly common practice of institutionalization emerged, which involved the creation and widespread adoption of "insane asylums." These institutions were essentially human warehouses, overcrowded and often deplorable in their brutal methods of operation. This only served to drive an increasing need to alleviate the suffering of those with debilitating disorders, as well as to relieve society of the financial burden caused by maintaining these institutions. And so, an idea emerged in the medical community: If the prefrontal cortex controls personality, perhaps by surgically damaging this brain region and its connections with the rest of the brain, we could transform personalities and assuage the troublesome behaviors associated with mental illnesses. Such a solution was exactly what psychiatrists in the 1930s were looking for—a bold pathway toward the radical metamorphosis of human behavior.

Interest in the prefrontal cortex had reached a pinnacle in 1935 when a neurology congress in London included a special meeting to discuss its function. Portuguese neurologist António Moniz attended this symposium and presented the idea that an ablation surgery on the prefrontal cortex may be a solution for treating a broad host of behaviors associated with mental illnesses. Soon after, on a fateful November 12, 1935, at a hospital in Lisbon, Moniz led a series of operations on patients suffering from psychiatric conditions of depression, schizophrenia, panic disorder, and mania. This launched a twenty-year period of the frontal lobotomy being considered an acceptable, although always controversial, medical procedure. Moniz's surgical series started with destructive alcohol injections into the prefrontal cortex and morphed into the use of specially designed blades to destroy networks between the prefrontal cortex and the rest of the brain. Despite a lack of clearly documented positive outcomes, this practice spread throughout Europe and the United States, where it was popularized by American neurologist Walter Freeman, who accelerated its adoption in 1946 with the "icepick" lobotomy. Freeman modified the procedure so that it could be performed in a physician's office, rather than an operating room. His transorbital approach involved lifting the upper eyelid, placing the point of thin, sharp instrument against the top of the eye socket, and then using a mallet to drive the tool through the bone into the brain. A prescribed series of cutting movements then effectively destroyed the connections entering and exiting the prefrontal cortex on one side of the brain, prior to the same procedure being repeated above the other eye, resulting in characteristic double black eyes. Reminiscent of Gage's tamping iron accident, these ice picks took a slow, methodical, but no less destructive path through the most evolved regions of the human brain. By the 1970s, approximately 40,000 individuals in the United States alone had been "lobotomized" before the procedure fell into disrepute and the practice was ceased.

Evidence that frontal lobotomies offered any real benefit to patients, or even to society, was thin at best, to say nothing of the ethical issues raised by its practice on individuals with personalities that were often merely deemed to be "difficult." What is clear now, however, is that lobotomized patients experienced similar changes in personality to those shown by Gage, which eventually led to the realization that the consequences of lobotomies were as bad, if not worse, than the conditions they were trying to treat. Even the physician who cared for Moniz's first lobotomy patients denounced the surgery, and described his patients as suffering a "degradation of personality." Walter Freeman offered a more detailed description of his patients that, as disturbing as it is, offers major insights into the complex role of the prefrontal cortex in behavior:

> Once the patient returns to his home, about two weeks after lobotomy, the newly emerging personality is fairly well developed and continues its evolution for a period of many months or even years. It is not a healthy personality at first; probably the word immature best describes it. Granting that the disease symptoms have been relieved by operation, the patient manifests two outstanding traits that could be described as laziness and tactlessness. In some people indolence is outstanding, in others hastiness, explosiveness, petulance, talkativeness, laughter, and other signs of lack of self-control. These patients know that they ought to busy themselves about the house, but they procrastinate; they know they should be considerate of their relatives and dignified in the presence of strangers, but it's too much trouble … their interest span is short, they are distractible.[10]

We share both the historical anecdotes of Gage and of lobotomized patients here not because they illustrate the role of the prefrontal cortex in personality, which is often the takeaway message, but because in these descriptions we find the first evidence of the role of the prefrontal cortex in cognitive control. A deeper look into Harlow's description reveals that Gage did not just become a jerk to his friends, but rather that the destruction of a large area of his prefrontal cortex led to a fundamental shift in the way he interacted with the world around him, as well as the world within him. This is clear from his tendency of "devising many plans of future operations, which are no sooner arranged than they are abandoned in turn for others appearing more feasible." The underlying commonality between the shift in how Gage treated other people and how he lived his life was that he lost something critical—*he lost control.*

We see similar evidence of loss of cognitive control in Freeman's descriptions: "These patients know that they ought to busy themselves about the house, but they procrastinate; they know they should be considerate of their relatives and dignified in the presence of strangers, but it's too much trouble … their interest span is short, they are distractible." Freeman goes on to describe that his lobotomy patients exhibit what is now clear to us to be degradation of the full extent of the cognitive control abilities that are necessary for them to enact their goals: attention, working memory, and goal management.

> The housewife complains of forgetfulness when she is really describing distractibility and lack of correct timing of the various household maneuvers. The man makes up his mind to look for a job but he can't quite summon up the energy necessary to overcome his inertia, and the many facets of the problem of obtaining employment are too numerous for his still limited capacity for consecutive and constructive thought. If he has a job to go back to he is apt to lose it because of errors of judgment and foresight. A lawyer noted that previous to operation he had been able, following an interruption, to resume his dictation at the exact word, whereas afterwards he had to have his secretary read her notes to him before he could resume his trend of thought.[11]

Damage to the prefrontal cortex turned Gage and these unfortunate recipients of frontal lobectomies into poster children for the Distracted

Mind. It is clear from these descriptions alone that the prefrontal cortex is critical for guiding our behaviors based on top-down goals, rather than responding reflexively in a bottom-up fashion to the world around us. This is true whether it is comes to withholding an inappropriate comment, or passing on a job opportunity because another one presents itself that captures attention in the moment. It is all about the loss of control. That is what our prefrontal cortex and its networks offer humans to differentiate us from much less evolved animals: the ability to pause in response to a stimulus and enact complex goals in a nonreflexive way. Harlow brilliantly intuited this solely based on observations of his patient Phineas Gage: "his equilibrium, or balance, so to speak, between his intellectual faculties and animal propensities seems to have been destroyed."

But it was not until the advent of sophisticated neuropsychological testing in the mid-twentieth century that our understanding of prefrontal cortex function expanded beyond its role in personality to include the cognitive abilities of executive function and cognitive control.[12] This includes the diverse set of operations involved in both *goal setting*, such as evaluation, reasoning, decision making, organization, and planning, and *goal enactment*, such as attention, working memory, and goal management. Evidence of prefrontal cortex involvement in these cognitive operations now includes findings from physiology and lesion studies performed on experimental animals and neuropsychology, electrophysiology, and functional brain imaging studies performed on humans. The transition of the prefrontal cortex from being a brain area shrouded in mystery, to being the source of our personalities, to being critical for establishing and enacting our goals was complete by the late twentieth century. Over the last three decades, extensive research efforts have been directed at elucidating the specific role of subregions of the prefrontal cortex, as well as network interactions between the prefrontal cortex and other brain areas, such as the parietal cortex, sensory cortex, and subcortical structures. However, many mysteries still exist, such as the neural basis of the dissociation in evolutionary development of our highly evolved goal-setting abilities and our ancient, limited, goal-enactment abilities, which generate the uniquely human Distracted Mind.

NEURAL NETWORKS

Understanding the role of the prefrontal cortex in the goal-enactment abilities of cognitive control is critical for our journey toward understanding the source of the Distracted Mind. However, these insights alone will not generate a complete picture. We now appreciate that the most authentic way to comprehend how the prefrontal cortex guides cognitive control is to focus not solely on the contribution of this brain region in isolation, but rather on how it interacts with other regions. Our closest ancestors, who also engage in complex interactions with their environments, may have a relatively smaller frontal lobe than ours, but certainly not dramatically so; it accounts for 36.7 percent of our entire brain volume, compared to modestly lower percentages in macaque monkeys (28.1 percent), gorillas (32.4 percent), and chimpanzees (35.9 percent). Recent research has revealed that what truly differentiates our frontal lobe from that of other animals is its extensive and complex interconnectedness with the rest of the brain via neural networks.[13]

Theories of brain organization have included two basic principles: *modularity*—the existence of neuronal assemblies, or island-like modules, with intrinsic functional specialization—and *neural networks*—the integration of information across distant brain regions. Both the modular and network principles of brain organization have a long and rich history.[14] The concept of modularity likely emerged at the end of the eighteenth century with the study of phrenology, a medical practice based on the theories of Franz Joseph Gall, a Viennese physician. Gall, seemingly inspired by his own childhood observations that the shape of his classmates' skulls and facial features were related to their cognitive abilities, came to believe that the brain exerted pressure on the skull to shape it. He reasoned that this pressure resulted in characteristic surface bumps that reflected the different functions of the brain structures residing underneath them. Gall went on to develop a systematic methodology to measure surface features of a person's skull to explain their cognitive strengths, weaknesses, and personality traits. Although we now know that this assertion is completely absurd, an important perspective from Gall's ideas was captured in a letter that he wrote to a censorship official in 1798: "The faculties and propensities of man have their seat in the brain ...

[and] are essentially distinct and independent: they ought, consequently, to have their seat in parts of the brain distinct and independent of each other."[15] Although the practice of phrenology has been discredited and now lives on only as a famous example of pseudoscience, the concept of functional localization in the brain that Gall described in his letter continues to thrive and accumulate empirical evidence. Phrenology died, but it gave birth to the modular view of brain organization.

In 1861, as phrenology was falling into disrepute, Paul Broca, a French physician and anatomist, localized critical aspects of language to the inferior region of the left frontal lobe by studying the brains of stroke patients who had lost their ability to communicate. This offered the world important anatomical evidence for the existence of the functional specialization of distinct brain regions.[16] Despite this evidence, not all scholars of the time supported this modular view of brain function. One of the earliest opponents of the localization perspective was a contemporary of Broca named Pierre Flourens, a French physiologist who argued that some forms of knowledge were not localizable, but rather were dispersed throughout the cortex of the brain.[17] He based this view on his research that involved ablating discrete regions of cortex in experimental animals and observing the impact on their behavior. His experimental results did support the conclusion that some aspects of brain function were indeed attributable to distinct brain regions, but he was unable to localize other "higher" cognitive abilities, such as memory. And so he presumed that these functions were distributed features of the brain, thus establishing early perspectives of what has become the network view of brain organization.[18]

Experimental support for the modular model of the brain continued with the development of neurophysiology, which characterizes the processing of single neurons and local collections of neurons, but also via neuroanatomy, which describes structural distinctions between brain regions. The field of neuropsychology supported modularity as well by demonstrating links between cases of isolated brain damage and specific behavioral and cognitive impairments. At the same time, momentum was building for the network model of brain organization based on anatomical studies that revealed extensive structural connections between widely distributed brain areas, and physiological studies that documented functional connections between distant regions in animals and functional imaging in humans.

Two centuries after Gall's letter, which presented what in retrospect is the basis of the modular model (minus the bumpy scalp nonsense), Joaquin Fuster, an American neuroscientist, outlined the tenets of a combined network and modular model, such that higher-order cognition emerges from complex neural networks that functionally connect distributed modules. He eloquently describes this model in his book *Cortex and Mind*:

> (1) cognitive information is represented in wide, overlapping, and interactive neuronal networks of the cerebral cortex; (2) such networks develop on a core of organized modules of elementary sensory and motor functions, to which they remain connected; (3) the cognitive code is a relational code, based on connectivity between discrete neuronal aggregates of the cortex (modules, assemblies, or network nodes); (4) the code's diversity and specificity derive from the myriad possibilities of combination of those neuronal aggregates between themselves; (5) any cortical neuron can be part of many networks, and thus of many percepts, memories, items of experience, or personal knowledge; (6) a network can serve several cognitive functions; and (7) cognitive functions consist of functional interactions within and between cortical networks.[19]

The majority of today's neuroscientists agree with Dr. Fuster's parsimonious perspective that does not disavow the existence of functional specialization and modules, but rather reconciles modular and network models of brain organization.

American neurologist Marsel Mesulam offered an important real-world example of how modules and networks are integrated. He showed that four different brain regions work together as nodes of a neural network to underlie the complex phenomena of selective spatial attention.[20] This is the same cognitive control ability that our ancestor used to search for the hidden jaguar. Dr. Mesulam described how brain lesions that were limited to any one of the regions in this network resulted in a partial attentional impairment, known as a neglect syndrome, while lesions involving all modules of the network resulted in a more extensive and debilitating attention deficit. And so, although the modules themselves do not completely embody the essence of attention, they serve as building blocks that underlie the complexity and diversity of this cognitive ability.

Thus, while the prefrontal cortex is a critical brain structure for our cognitive control abilities, the mechanisms that yield these abilities are not solely localized in the prefrontal cortex. Rather, they emerge from interactions across neural networks between the prefrontal cortex and regions throughout the brain. These prefrontal cortex networks direct cognitive control influences on our *sensory input, internal states*, and *motor output.*[21] The influence on sensory input involves the modulation of "representations" in the cortex, which are symbolic codes of information. This modulation occurs for all sensory modalities: visual, auditory, tactile, and olfactory. Influence on internal states includes the modulation of our emotions, thoughts, mental imagery, and internal voice. Influence on motor output involves the modulation of our bodily movements and other complex actions, such as speech.

Extensive connections stretch in both directions between the prefrontal cortex and other brain regions. This enables goal-directed control of the diverse information-processing systems that underlie all aspects of cognition. It is important to recognize that neural networks are not nebulous webs of connections that are engaged equivalently during all mental processes. On the contrary, they are precise and distinct. It is this specificity of neural networks that underlies the diversity of prefrontal cortex functions. For example, a specific area in the prefrontal cortex, the orbitofrontal region (lower region of the prefrontal cortex), is extensively interconnected with regions involved in controlling autonomic responses and emotion—the hypothalamus and amygdala, respectively. It is these connections that underlie the role of the orbitofrontal cortex in influencing our emotions based on our goals. When experimental animals are electrically stimulated in this region of the prefrontal cortex, there are effects on their respiratory rate, heart rate, blood pressure, and stomach secretions. Other control pathways that influence fear processing are mediated by networks that also emanate from this region of the prefrontal cortex.

However, these networks are not engaged in isolation. They interact simultaneously with many other prefrontal cortex networks, such as those interconnected with sensory regions, which modulate how perceptual information is processed based on goals. And, it is even more complicated than this, as the local neurochemical environments of the interconnected brain regions influence the functionality of neural networks. For example, the many neurotransmitter systems—such as dopamine, norepinephrine, serotonin, and acetylcholine—regulate prefrontal cortex function and thus the function of the areas to which they are connected. And it is more complicated still, as we have now come to understand that communication between brain regions is influenced by the synchronicity of the rhythmic patterns of neural activity in the different regions. It is beyond the scope of this book to detail the interactions between all of these concurrent and integrated processes, but it should come as no surprise that studying how prefrontal cortex networks underlie cognitive control has consumed decades of multidisciplinary scientific effort and we still have only scratched the surface.

TOP-DOWN MODULATION

It is widely accepted that the prefrontal cortex enables cognitive control by modulating neural activity in distant brain regions via long-range connections, or neural networks. This immensely important mechanism is known as *top-down modulation*. Studies from the Gazzaley Lab, as well as many others, have shown that it involves the modulation of both the magnitude of neural activity as well as the speed of neural processing within brain regions to which the prefrontal cortex are connected, based on an individual's goals.[22] This is the neural basis of how our goals bias the information we process.

As described by neuroscientists Earl Miller and Jonathan Cohen in their influential article, "An Integrative Theory of Prefrontal Cortex Function": "Depending on their target of influence, representations in the prefrontal cortex can function variously as attentional templates, rules, or goals by providing top-down bias signals to other parts of the brain that guide the flow of activity along the pathways needed to perform a task."[23] Miller and Cohen further posited that "cognitive control stems from the active maintenance of patterns of activity in the prefrontal cortex that represents goals and means to achieve them. They provide bias signals to other brain structures whose net effect is to guide the flow of activity along neural pathways that establish the proper mappings between inputs, internal states, and outputs needed to perform a given task." Thus, the cognitive control needed to enact our goals is manifested by higher-order representations in the prefrontal cortex that result in the top-down modulation of neural activity in other brain regions via widely distributed neural networks.

Top-down modulation as the fundamental mechanism by which the prefrontal cortex mediates cognitive control is consistent with what we have already discussed in terms of overall brain organization. As described in chapter 2, the front half of the brain is responsible for action, while the back half of the brain is specialized for perception. But the actions driven by the front part of the brain are not uniform; rather, they are organized in a hierarchical gradient, starting with the most primitive brain regions of the frontal lobe located in the motor cortex, and sweeping forward toward the most evolved structures in the prefrontal cortex. Action mediated by the motor cortex is very straightforward; it involves movements of our muscles, which is what most of us think of when we hear the term "action." Here, top-down modulation is mediated by projection neurons in the motor cortex to the spinal cord where signals modulate activity that results in muscle movements based on our goals.

As you advance forward along the frontal lobes, actions become more complex—for example, language expression and the intricate control of our eye movements. And even further forward in the prefrontal cortex, the concept of "actions" becomes yet higher level and more abstract. These actions and the top-down modulation that mediates them do not even leave the brain to result in observable events. For example, the prefrontal cortex sends projections to the visual cortex in the back of the brain, where it results in top-down modulation of visual cortex activity. This is the neural basis of selective attention, as will be described in more detail later in this chapter. You can think of all aspects of cognitive control as the higher-order actions of the prefrontal cortex that mediate our goal-directed control over how we both perceive and act in the world around us.

Note that the mechanisms of top-down modulation of visual attention and of body movements are very similar. Neurons in the motor cortex project to neurons in the spinal cord to modulate their activity and influence how we move in the world, while neurons in the prefrontal cortex send signals to neurons in the visual cortex to modulate their activity and influence how we perceive the world. These distinct networks originate from the different regions of the frontal cortex and project to different brain regions to influence movement and perception based on our goals, but they both do so via top-down modulation of neural activity. From an evolutionary perspective, top-down modulation by the prefrontal cortex of the activity in the sensory cortex to control attention to our sensory inputs, is a natural extension of the same mechanism used by more primitive regions of the frontal lobe to control simple movements. In this manner, the actions enabled by the prefrontal cortex via top-down modulation include all of cognitive control—attention, working memory, and goal management. Let's now explore each domain and show how prefrontal cortex networks and top-down modulation enable them.

ATTENTION

Our ancestor roams through the forest and comes upon a stream. To accomplish his goal of carefully assessing whether a jaguar is hidden nearby, he engages his selective attention processes to focus his vision on searching for a stippled orange and black pattern and directs his sense of smell to detect a musky odor. He fires both of these attentional arrows in the direction of the thick brush to the left of the steam.

How does our ancestor engage his attention in this way? As described, the principal mechanism that underlies cognitive control in general, and selective attention in particular, is prefrontal-cortex-mediated top-down modulation of neural activity. This occurs via engagement of long-range neural networks that connect the prefrontal cortex with other brain regions—in the described scenario, with sensory brain areas. This modulation biases neural activity patterns so that the selected stimuli that are most relevant to our ancestor's goals are represented more strongly in the brain regions that encode them—a stippled orange and black pattern in the right visual cortex and a musky smell in the olfactory cortex.[24] This biasing based on expectations results in neural representations of these stimuli having greater contrast against background activity, and thus they will be easier to detect if they are present. So, when a musky smell does enter his nostrils, the neural representations that are specific to these stimuli in his olfactory cortex are exaggerated and more obvious to him than if he had not set this as a goal. His goals have biased his perception. And so, selective attention enables him to act more effectively upon subtle information.

Although this description of how top-down modulation underlies attention is accurate, it does not tell the full story of how selectivity is instantiated in our brain. Modulation does generate a contrast of relevant representations against background activity, but this is not accomplished solely by enhancing of representations of the relevant signals. It also involves the suppression of representations of irrelevant information. This process serves to create greater contrast in the brain, allowing relevant signals to become even more salient. Just imagine how much "higher" you would jump off the ground if, when you leaped up, the ground also dropped down. Suppression of irrelevant information improves our focus on the relevant by dropping the floor of the irrelevant.

From an experiential viewpoint, the neural process of suppression can be thought of as the act of ignoring, which is a far more important action than many people realize. Decades of research have now shown that while focusing on relevant information is of course critical to accomplishing our goals, ignoring irrelevant information is just as important. This does not necessarily have to be a conscious process. Our ancestor likely was not trying to ignore the sounds of rodents scurrying along the forest floor and birds flittering through the trees because he understood that this would help him hear the jaguar's snarl. But, if you sit in a coffee shop and try to focus on a conversation with a friend, you indeed may be aware of the goal interference all around you and thus consciously attempt to ignore the chatter and café music. In either case, suppression of irrelevant information in your brain is not a passive process, but an active one that generates contrast between neural patterns and thus finely sculpts how we experience the world based on our goals.

In a study performed in the Gazzaley Lab, we showed young adults a series of two faces and two nature scenes presented one at a time in a randomized order (we will refer to this as the face/scene experiment throughout the book).[25] We told them which stimuli were relevant and should be remembered for a short period of time (seven seconds), and which stimuli were irrelevant and should be ignored. For example, in one part of the experiment we told them to remember the scenes and ignore faces and that we would test their memory for those scenes in seven seconds. In another part we told them to ignore the scenes and remember the faces. While they performed these tasks, we scanned their brain activity in an MRI scanner and used functional MRI (fMRI) sequences to assess the magnitude of brain activity in their visual cortex. We were interested in comparing the brain activity for stimuli when they were attended versus those same stimuli when they were ignored. We also compared activity levels to the activity for those same stimuli when they were viewed passively—that is, there was no goal to remember or ignore them. We found that there was more activity when our participants were attending to scenes than when they were passively viewing them, which we labeled *enhancement*—a neural measure of focusing—and that there was less activity when they ignored the scenes than when they passively viewed them, which we referred to as *suppression*—a neural measure of ignoring. What we learned from this experiment was that the act of ignoring is not a passive process; rather, the goal to ignore something is an active one that is mediated by the top-down suppression of activity below baseline levels of passively viewing. *The fact that ignoring is an active process is critical to understanding the Distracted Mind because it emphasizes that it takes resources to filter out what is irrelevant.*

In another study, we used electroencephalographic (EEG) recordings of electrical brain activity during this same task and found that participants processed relevant stimuli faster in the visual cortex than passively viewed stimuli and that they processed irrelevant information more slowly. From this experiment we learned that top-down modulation involves an influence on both the magnitude and speed of neural processing: there is greater and faster neural processing when attending, and lesser and slower processing when ignoring.

Suppression yields higher-quality representations of relevant information by decreasing the noise and letting the signal shine. It is an essential ingredient of selective attention. Although it may seem counterintuitive, we now appreciate that focusing and ignoring are *not* two sides of the same coin. In other words, it is not necessarily true that when you focus more on something, you automatically ignore everything else better. We have shown in our lab that different prefrontal cortex networks are engaged when we focus compared to when we ignore the same thing. In other words, they are two separate coins.[26] This means that your goal of focusing on a conversation in a restaurant may be successful, but your ability to ignore the chatter all around you may be failing. If so, you will find yourself susceptible to one of the two types of goal interference: distraction.

Until recently, most evidence that prefrontal cortex networks mediate selective attention via top-down modulation was based largely on recordings of brain activity that showed both the activation of the prefrontal cortex and modulation of activity in the sensory cortex during an attention-demanding task. Scientists have observed this finding with many different techniques. Dr. Gazzaley's own research took this a step further by using fMRI data and an analytical approach known as functional connectivity that permits the study of neural networks. This approach involves calculating correlations in activity patterns between different brain regions across multiple trials of a task.[27] It is based on the idea that if two brain regions exhibit the same pattern of activity—that is, they show similar fluctuations (ups and downs) in neural activity over many repetitions of the same task—then they are likely to be within the same neural network. We showed in our study that a region in the prefrontal cortex was functionally connected with a region in the visual cortex, thus defining them as nodes in a network.[28] Importantly, we showed that the magnitude of the functional connectivity between these brain regions depended on the relevance of that information to the participant's goals. Moreover, we found that the strength of the connection between the prefrontal cortex and visual cortex correlated with the magnitude of the enhancement and the suppression of visual activity. These results suggested that the prefrontal cortex drives top-down modulation of activity levels in the visual cortex by modifying the strength of the neural networks that connect these regions in accordance with the goals of the task.

However, the functional connectivity technique we used is still a correlational approach—that is, it still does not offer causal evidence that the prefrontal cortex mediates top-down modulation via neural networks. These data suggest that the prefrontal cortex is involved, but offer no proof that it is required for top-down modulation. An optimal experimental design to assess causality is to disturb the function of the prefrontal cortex while someone is engaged in a selective attention task and simultaneously record neural activity in functionally connected regions of the visual cortex. This would allow us to determine whether top-down modulation in the visual cortex is disrupted when the prefrontal cortex function is perturbed, and thus show that they are in fact causally related.

The Gazzaley Lab recently conducted this very experiment, in a study lead by Dr. Theodore Zanto.[29] First, our research participants came to the lab for an MRI scan while they performed a visual attention task. We analyzed their data using our functional connectivity approach to locate regions in their prefrontal cortex that were potential nodes of an attention network with their visual cortex. Then, on another day, each participant returned to the lab and received repetitive magnetic pulses directed to this region of their prefrontal cortex area using transcranial magnetic stimulation (TMS). This use of repetitive magnetic pulses to the scalp has been shown to be a safe way to temporarily disrupt the function of the underlying cortex for a brief period of time after it is applied. Then, immediately after the TMS was delivered, our research participants performed the same visual attention task that they engaged in while they were in MRI scanner, but this time with EEG recordings. We found that disrupting function in the prefrontal cortex with TMS diminished top-down modulation of activity across the brain in the visual cortex and also decreased participants' ability to remember the relevant information a brief time later. This experiment thus generated important evidence that the prefrontal cortex *causally* induces top-down modulation of activity in the visual cortex and that this activity modulation (both enhancement and suppression) underlies selective attention that is necessary for working memory performance. Thus this study also advanced our understanding that aspects of cognitive control are intimately related. But as we will see now, these aspects do also have several distinct mechanisms.

WORKING MEMORY

After spotting the stream, our ancestor darts for cover and holds visual details of the scene he just saw in his mind—for example, where the brush was located in relation to the steam. He actively maintains this information in mind until he feels it is safe enough to poke his head out and begin his search for the jaguar.

Understanding the neural basis of working memory has been one of the greatest challenges in the field of cognitive neuroscience. Maintaining information in our minds after it is no longer present in the environment is critical for all higher-order behavior. However, the mechanism of this phenomenon turned out to be very elusive to pin down. Dr. Carlyle Jacobsen and colleagues were the first to establish the critical role that the prefrontal cortex played in working memory in the 1930s. They showed that deficits in working memory performance could be induced by experimental lesions to the prefrontal cortex of monkeys using a task where information is presented, held in mind throughout a delay period, and then probed for recall.[30] But it was not until 1971 that we made another major leap in understanding the neural basis of working memory. In that year Dr. Joaquin Fuster and colleagues reported their discovery of neurons in the prefrontal cortex of monkeys that showed persistent neural activity after a stimulus was no longer present.[31] The monkey's neural activity persisted for a relevant stimulus and remained elevated throughout the time when the monkey was holding relevant information in mind. These remarkable neurons were termed "memory cells." Persistent brain activity in the absence of visual stimulation, further characterized by the seminal work Dr. Patricia Goldman Racik, came to be viewed as the neural signature of maintaining information in working memory.[32]

Over the years, other studies have revealed that neurons across the brain exhibit this same property of persistent activity after a stimulus is no longer present. Notably, this has been observed throughout sensory cortex, where stimulus features were being represented in the brain. In fact, persistent activity in the sensory cortex turns out to be another example of top-down modulation, similar to what occurs during selective attention, but in this case, it happens when the stimulus is no longer present. Our current understanding is that the prefrontal cortex mediates the modulation of activity in the sensory cortex during working memory using the same networks that are used for selective attention when a stimulus is present in the environment. This is how we keep information alive in our minds when it is no longer right in front of us. Fuster also contributed causal evidence by showing that disrupting the function of the prefrontal cortex in monkeys by reversibly cooling this brain area affected the modulation of activity in the visual cortex, as well as impaired working memory performance.[33]

GOAL MANAGEMENT

While he unsuccessfully searches for the jaguar, our ancestor decides to throw a stone into the water to induce some movement. And so, after he ducks for cover, he searches for the perfect stone, all the while continuing his original goal of jaguar searching, as well as maintaining a detailed visual image of the stream and brush on the left bank.

Having goals that require accomplishing more than one task at a time is an aspect of our behavior that stretches back to our distant ancestors. This behavior is often referred to as "multitasking," a term that was borrowed from computer science where it connotes the parallel processing of information. But what actually occurs in our brains when we engage in multitasking behavior? Do we truly "parallel process"? Well, here the devil is in the details. Our brains certainly parallel process plenty of information. We are constantly receiving extensive data from our sensory system that is processed subconsciously, not to mention all the processes that are always engaged to maintain our breathing, heart rate, and so on. Even when it comes to action, parallel processing abounds. If one or more tasks are capable of being automated as reflexes, then they can easily be engaged simultaneously with another task without much consequence. This is the classic "walking while chewing gum." Although the act of walking requires selective attention, the act of chewing under most circumstances does not demand cognitive control because it is performed reflexively. Given this, such an activity may not even qualify as an example of multitasking, since a reflexive action is not really a task. But if the two goals both require cognitive control to enact them, such as holding the details of a complex scene in mind (working memory) at the same time as searching the ground for a rock (selective attention), then they will certainly compete for limited prefrontal cortex resources.

In another study performed in the Gazzaley Lab, we investigated what happens in the brain when participants performed two cognitive control-demanding tasks simultaneously, similar to those that were engaged by our ancestor in our scenario.[34] To do this, we instructed our research participants to pay attention to a nature scene presented to them on a computer screen, and hold the details in mind over a seven-second delay period, after which they were tested on how well they remembered it. The twist here was that on some trials of the task they had to perform another task at the same time. While they were holding the scene details in mind, they had to make an age and gender decision about a face that flashed on the screen during the delay period. This secondary task required selective attention and thus led to competition with the resources already online for the working memory challenge. This is directly analogous to our ancestor making a decision about which stone to pick up while holding details of the scene around him in mind.

What we found in this experiment is that, as expected, participants' prefrontal cortex network is engaged when they view the nature scene, and this network activation persists into the delay period. This working memory maintenance network is responsible for retaining the image of the scene in mind. As described previously, it involves functional connectivity between the prefrontal cortex and visual cortex that drives the top-down modulation of activity in the visual areas that represent the scene. But we also found that when the face was flashed on the screen during the delay period and a decision was made, the working memory network diminished, as well as the visual activity that was involved in maintaining the scene in mind. At the same time, a new network between the prefrontal cortex and visual areas became engaged to represent the face, which should not be surprising as this is the mechanism of selective attention. What is telling is that both networks were not engaged equivalently at the same time. And after the face was gone from the screen, we could see the face attention network diminish and the original working memory network for the scene become reactivated, corresponding to the participant's expectation of being imminently tested on his or her memory of the scene.

Even though we gave no instruction to participants to switch between these two tasks, we see that is actually what is happening in their brain. They do not maintain the memory network at the same level when the selective attention network is engaged. Rather, they dynamically switch between these two cognitive control networks. The results of our study are consistent with many other studies that have shown when we simultaneously pursue multiple goals that compete for cognitive control resources, our brains switch between tasks—they do not parallel process.[35] So, while the behavioral goals may have been to multitask (and thus "multitasking" is an appropriate term for this as a behavior), the brain itself is *network switching*. As we will discuss in the next chapter, this act of switching, whether we make the decision to switch or not, diminishes our performance on tasks. This is the basis for the other type of goal interference: Interruption.

Although the special role of the prefrontal cortex in cognitive control is clear, to say it is the only player in cognitive control would be a large oversimplification. The mechanisms of cognitive control involve a broad network that includes many other brain regions, such as the premotor cortex, parietal cortex, visual cortex, and subcortical regions such as the caudate, thalamus, and hippocampus. For example, it was demonstrated that the volume of a region in the parietal cortex predicted individual variability on a self-report measure of everyday distractibility.[36] A detailed discussion of the contributory role of all of these brain regions in cognitive control is beyond the scope of this book, but the takeaway message is that cognitive control is mediated by top-down modulation and the coordinated functional interactions between many nodes of widely distributed and interacting networks in the brain. With that background, let's now discuss the distinct limitations that exist for all of our cognitive control abilities and how this leads to the Distracted Mind.

4 CONTROL LIMITATIONS

IT IS NOW well understood by neuroscientists that our cognitive control abilities are far from ideal. Each of the components—attention, working memory, and goal management—have deeply embedded functional limitations that result in suboptimal performance as we attempt to accomplish our goals. This is especially true when our goals lead us to engage in interference-inducing behaviors—multitasking in distracting settings—which is now commonplace in our high-tech world. A major premise of this book is that our Distracted Minds are generated by a head-on collision between our high-level goals and our intrinsic cognitive control limitations; this conflict generates goal interference that negatively impacts quality of life. If we want to overcome the derailing forces of goal interference, we need to increase our understanding and expand our awareness of our cognitive control limitations so that we can find ways to minimize those factors.

The tools of human neuroscience have been incredibly valuable in enlightening us on both how cognitive control allows us to accomplish our goals and the neural basis of its limitations. This offers us the opportunity to better understand the underpinnings of our modern interference dilemma. Let's tour the limitations in cognitive control that are at the core of the vulnerabilities in our brain's information processing systems that give birth to the Distracted Mind.

ATTENTION LIMITATIONS

Selectivity

Engaging our attention with the highest level of selectivity is critical for us to function effectively in the complex environments that we inhabit. Our brain simply does not have the infinite parallel processing resources needed to simultaneously receive and interpret all the information we are exposed to at every moment. And so, we need to rapidly fire our cognitive resources at targets selected to be most relevant to our goals. Simultaneously, we must block out the vast, rapidly changing stream of goal-irrelevant information that flows around us.

Of course, this necessity to focus and ignore is not new to humans living in modern technologically rich societies. Even during the early days of brain evolution, prior to the development of neural mechanisms of goal-directed attention, information processing needed to be selective. The inherent limitation in a brain's parallel processing capacity is likely what drove the earliest evolution of our selectivity mechanisms. Such limitations permit only the most novel and salient events and objects in the environment—notably those offering survival and reproductive advantages—to generate the strongest representations in our brain and thus exert the greatest influence on our perceptions and actions. As described, this earliest form of selectivity is known as bottom-up processing, sometimes considered itself to be a form of attention—although not one based on top-down goals, but rather driven by stimulus properties themselves. This ancient form of attention served as the earliest fuel of the perception-action cycle, and it remains deeply rooted in our modern brains.

Bottom-up sensitivity persists as an essential asset for the survival of all animals, including humans. It is easy to imagine that our lives, either in the city or the wilderness, would not last as long if we did not retain the ability to rapidly and automatically detect warning signals from the environment: the blare of a car horn as we carelessly step out into the street, or the sound of a falling rock as we stroll unwittingly along a forest trail. Bottom-up sensitivity is critical, especially if our top-down goals are directing our attention down a focused path. Thus, driven by the evolutionary forces of natural selection, this primitive influence has remained a core component of how we interact with the world around us.

All our interactions with the environment involve a constant, dynamic integration between these two great modulators: top-down attention and bottom-up processing. Our brains are constantly having a tug of war between these two forces, with the winner of the contest exerting the strongest influence on our perceptions and actions, which cascades into a direct impact on our behavior.

Despite its necessity, when attentional selectivity is viewed from the perspective of being a core cognitive control ability that enables us to enact our goals, retained sensitivity to bottom-up influences represents a serious limitation. It is a remnant of our ancient brains and a major challenge to any selectivity mechanism whose core function is to filter out everything that is not in the crosshairs of our goals. Those stimuli with the strongest bottom-up factors of novelty and salience are the most formidable at involuntarily usurping our attention from our goals; they are the source of external distraction and a major aspect of goal interference.

A rich literature from the fields of psychology and neuroscience describes the source of our limitations in attentional selectivity. One prominent series of findings that has shaped our current thinking is that selectivity depends as much on neural processes involved in ignoring goal-irrelevant information as it does on processes that facilitate the focus on goal-relevant information. This relationship between focus and ignore has been described as a "biased competition," or a push-pull battle between top-down and bottom-up processes.[1] Neural data show that when two objects are simultaneously placed in view, focusing attention on one pulls visual processing resources away from the other. But when the stimulus sitting outside of the area of our goals has intrinsic characteristics that induce strong bottom-up attention, then the competition is not so easily won by our goals. Just imagine focusing your attention selectively on a conversation you are having in a noisy restaurant. All the while, even despite your lack of awareness, an internal competition is raging to resist the source of goal interference. Then an altercation erupts at the next table. Undoubtedly, this bottom-up influence will win the attention battle despite your goals of ignoring everything except the conversation. Our attentional selectivity is limited; we don't always get to fire our cognitive arrows precisely where we want without interference.

The Gazzaley Lab studied the neural basis and the consequences of limitations in attentional selectivity by asking healthy young adults to focus their attention on trying to remember the color of a field of stationary dots for a brief time.[2] The real challenge was that every second or so the dots would lose their color and all start to move at the same time in one direction. Our participants were clear that their goal was to maintain their focus solely on remembering the color of the stationary dots and ignore the movement of the dots. Although overall they did quite well on this simple working memory test, we found that they did not perform equivalently well on every trial; on some trials they took longer to arrive at the answer as to the correct color of the dots.

While they performed this task we recorded their brain activity using EEG. After analyzing their activity patterns we discovered that their memory was better on faster trials not because those were the trials where they exhibited the best focus on the colored dots, but because they were trials where they were best at ignoring the moving dots. This experiment revealed that focus was not the primary determinant of high-level working memory performance; rather, memory depended more on effectively ignoring distractions. We generalized the conclusions from this study by showing the same exact findings when the instructions were reversed and participants had to remember the direction of motion of the dots and ignore the color of the stationary dots. Note that both motion and color are notoriously strong bottom-up influences. Here we learned that our ability to ignore goal-irrelevant information is fragile, even in healthy twenty-year-old brains. Moreover, we showed that a failure to ignore information results in overrepresenting the distracting piece of information, which in turn interferes with maintaining the relevant representations in mind, which leads to diminished success in goal-directed behavior.

As other research groups have shown, failure to effectively ignore irrelevant information has direct consequences for our success at holding relevant information in mind for brief periods of time.[3] But what about long-term memories? Do limitations in selective attention affect them as well? We addressed this question in another experiment that was led by Dr. Peter Wais in the Gazzaley Lab. It was in part inspired by the common observation that when someone is asked to recall the details of a past event from memory they frequently look away, or even close their eyes as they prepare to answer. Go ahead and try it: ask a friend to recall in detail what he had for dinner last night. Make sure you watch his eyes carefully. You will likely see that he looks away from you before he answers. This tendency to look away has actually been associated with better memory recall.[4] We hypothesized that the reason was because the mere act of looking at a face while probing memories is distracting and interferes with the directing of selective attention internally that is needed to recall the details of the past event.

In the experiment, we first asked our research participants to answer questions about 168 novel images presented to them on a computer monitor. The images contained between one and four items of the same object, for example, a single book or four books. The questions were: "Can you carry these object(s)?" and "Can you fit them inside a woman's shoebox?" Unbeknownst to them, this was the study phase of a memory test. One hour after they viewed these images, they climbed inside the MRI scanner where they were asked one at a time how many items were present in each of those images. While this went on we recorded both their memory performance and what was going on in their brains. The twist was that they performed the memory test either with their eyes closed, their eyes open looking at a gray screen, or their eyes open looking at a picture, thus mimicking the complex visual world that is usually in front of our eyes when we recall things from memory. The pictures they saw during the memory test were entirely irrelevant distractors and they were specifically instructed to ignore them.

The results of this experiment revealed that their ability to remember details, as indicated by an accurate report of the number of objects in the previously viewed image, was significantly diminished when their eyes were open and there was a picture in front of them, compared to either their eyes being shut, or their eyes being open while they faced a gray screen.[5] These results suggest that the presence of the bottom-up distraction of a complex picture diminished the participants' attention needed for searching their memory. It was not just having their eyes open, because looking at the gray screen revealed the same memory quality as having their eyes closed. It was the presence of the busy scene that resulted in the distraction effect.

This finding was similar to the results of the "dots experiment" in that even when our participants were told explicitly to ignore irrelevant visual information in front of them, they frequently failed to do so, which resulted in deficient memories. The fMRI results from this experiment further revealed that diminished memory recall in the presence of a distracting scene was associated with a disruption of a neural network involving the prefrontal cortex, hippocampus (a brain area involved in memory consolidation), and the visual cortex. The results led us to conclude that bottom-up influences from viewing irrelevant pictures led to interference with our participants' top-down goals of recalling the number of objects in the images. We also learned that prefrontal cortex networks, which serve a critical role in selective attention, are quite vulnerable to being disrupted by these low-level bottom-up influences.

The conclusion of our fMRI study was that disruption of prefrontal cortical networks underlies limitations in selectivity that result in memory deficits. However, it is important to appreciate that these results do not permit us to make causal conclusions about the role of the prefrontal cortex in fending off distraction. We cannot tell for sure that disruption of these networks would worsen the hallmarks of a Distracted Mind; at this point we knew only that they were associated with it. To learn more, Dr. Wais performed a follow-up study to disrupt function in the prefrontal cortex by first applying repetitive transcranial magnetic stimulation (TMS), which temporarily disrupts brain function in that area, and then recording the impact on performance in the same long-term memory experiment. The results showed that the negative impact of passively viewing the pictures on memory recall was even *worse* if we disrupted prefrontal cortex function, thus supporting our hypothesis that prefrontal cortex networks support memory recall by reducing distractions.[6]

In a related experiment, we were interested in determining if the distraction effect on long-term memory was driven by the common visual nature of the distracting picture and the visual memory that the participants were attempting to recall (the number of items in an image). To assess this, a new group of participants performed exactly the same long-term memory test, except here they always kept their eyes open while viewing a gray screen. The twist was that they answered the memory questions either in silence, while white noise played, or while they heard busy restaurant chatter that they were instructed to ignore. We found that their ability to recall the details of the *visual* memories were just as diminished by the auditory distractions (restaurant chatter) as they were by the visual distractions (busy picture).[7]

To be clear, these results are not meant to serve as the basis for advice to walk around with eye masks on and earplugs inserted. They are shared to illustrate the surprisingly high degree of sensitivity our attentional selectivity filter has to distraction and how it negatively affects the recall of our long-term memories. The results contribute to our basic understanding of the neural basis of limitations in attention, which causes seemingly innocuous acts—like having our eyes and ears exposed to everyday stimuli—to diminish our ability to recall the details of our memories.

The reality is that eye masks and earplugs only help so much. Another factor in why our cognitive arrows miss their targets is the presence of internally generated distraction, or mind wandering. A clever research study used an iPhone app to randomly present questions to college students asking if at that very moment they were focusing their attention on what they were doing or if their mind was wandering. Strikingly, the study revealed that 47 percent of randomly sampled moments throughout the day were occurrences of mind wandering.[8] In addition, they found that people were generally less happy while mind wandering, seemingly independent of the type of activity that they were engaged in at the time. Mind wandering has been shown to have a negative impact on cognitive performance, and it has been associated with deficits in working memory, fluid intelligence, and SAT performance.[9]

Although mind wandering is often quite benign, it can also be very disruptive when a task demands high-level performance, such as during a critical meeting or when driving in traffic. At an extreme it can be so disabling that it is associated with psychiatric conditions such as major depression, post-traumatic stress disorder (PTSD), and obsessive-compulsive disorder (OCD). For these unfortunate individuals, internal distraction essentially shuts down goal-directed behavior, resulting in debilitating impairment to their functioning.

Externally generated, bottom-up influences by sights and sounds and internally generated mind wandering both blunt the sharpness of our attentional selectivity. While it is clear that limitations in the selectivity of attention are a major factor in contributing to the Distracted Mind, this is not the whole story. We display limitations in all other aspect of our attention as well: our ability to distribute attention broadly, to sustain attention for long periods of time, and to deploy attention rapidly. Let us explore each of these.

Distribution

When it comes to attention, we do not always want to shoot it like an arrow; sometimes we want to do the opposite and distribute our cognitive resources as broadly as possible, like a fisherman casting a wide net into the sea. The major factor that guides us in reaching a decision about which of these two methods of deploying our attention we choose is the level of predictive information we have prior to an event. To extend the fishing analogy, predictive knowledge of exactly where the fish are swimming helps to determine whether you want to use a spear or a net. Imagine again our ancestor who did not know exactly where the jaguar might be lurking, but only that it was likely hidden somewhere to his left side. In that case, he would want to direct his attention broadly to the left, rather than directing it precisely at the brush on the bank of the stream. With less predictive information, a laser beam of attention applied to the wrong place would be a suboptimal strategy. In fact, it would result in suppressing the detection of a jaguar at all other locations. Distributing attention is what we engage in when we have less precise information. An everyday occurrence of distributed attention takes place when you drive a car. In this situation, you need both selective and distributed attention—to hold your focus on the road while also maintaining sensitivity to unexpected events in the periphery, such as someone carelessly stepping into the street while he talks on his mobile phone.

Expectations are fueled by predictive information about where, when, and what events will occur in the immediate future. Less detailed predictions about future events lead us to distribute attention rather than focus it. But distributed attention is still selective in many ways. In the scenario where the jaguar might be anywhere to the left side, our ancestor continues to listen, watch, and smell for the jaguar-specific features; he just does it over a wider area of space. This act of distributing attention can apply to any sensory domain; for example, he might not have known *exactly* how a jaguar smells, but he knows that it has a musky odor, and so he distributes his olfactory attention to encompass general musky odors.

The major limitation is that when we distribute our attention broadly, the benefits we gain from selective attention are diminished. This was demonstrated recently in the Gazzaley Lab by asking participants to hold their eyes positioned on the center of a screen while we flashed them a cue that gave varying degrees of predictive information about where an upcoming target would appear in their periphery: a 100 percent cue told them exactly where it would appear, a 50 percent cue told them which side it would appear on, left or right, and a 0 percent cue gave them no predictive spatial information. When the target appeared seconds later they were both faster and more accurate in distinguishing between it and a distractor when they were given a 100 percent cue versus a 50 percent cue. They were even slower and less accurate when they had no information as to where it would appear.[10] This shows that we do not gain the same benefits of attention when we distribute our resources compared to when we have the predictive information needed to be selective.

Translating this experimental finding to our ancestor's situation would reveal that he had the best chance of detecting the jaguar if he had more information about where it might be hidden. If there were only one bush behind which a jaguar might hide, then he could selectively focus his attention there, while if there were five bushes he would have to distribute his attentional resources over all five, an act that would diminish his chances of spotting the lurking predator. In general, our ability to distribute our attention is quite limited.

Sustainability

In addition to limitations on our selectivity and distribution of attention, we also have limits on how well we can sustain our attention over time, notably in non-engaging—and especially in boring—situations.[11] Sustained attention, sometimes referred to as vigilance or attention span, is most frequently assessed by measuring how well someone maintains consistently high-level performance on a repetitive task over a long period of time. Just imagine our ancestor crouching expectantly behind the tree, watching ... listening ... smelling ... for signs of a jaguar. How long would he be able to sustain his attention at a highly selective level with minimal or no feedback? The consequences of missing subtle cues could be life ending. If his mind wandered for a moment or he became impatient after only a few seconds and approached the stream, he likely would be the jaguar's lunch.

Most research studies on the limits of sustained attention use very boring tasks, which are directed at assessing vigilance in the context of low levels of stimulation. This is the ability that aircraft traffic controllers are expected to bring on in spades to perform their jobs. But clearly there is much more to sustained attention than this; for example, why does it vary so much between individuals? This knowledge is critical for expanding our understanding of the Distracted Mind, both for people performing at a high level and those who have been diagnosed as having an attention deficit. The need to expand our understanding of the context-dependent nature of limitation in attention span, and indeed for all cognitive control limitations, is made clear by a recent study that showed that children diagnosed with ADHD (attention deficit hyperactivity disorder) had difficulties sustaining attention when they were assessed using standard boring lab tests, but not when playing engaging video games.[12] Parents are often amazed that their children with ADHD cannot stay focused on their homework for more than a few minutes but can play a video game for hours on end.

Speed

The final limitation of our attention that we will discuss is processing speed. Although each neuron in our brain performs computations at an incredibly rapid pace, clocked in the thousandths of a second (milliseconds), attention, like all aspects of cognitive control, is an emergent property of neural networks that relies on the integration between signals from multiple regions distributed throughout the brain. This transfer of information, much of which is serial in nature, invariably introduces significant time delays at each relay in the network, resulting in aspects of attentional processing taking tenths of a second. While this may still sound fast, it is actually slow given how rapidly our interactions with the environment take place.

Researchers have studied the processing speed limitations of attention in experiments using a paradigm called the "attentional blink."[13] In these experiments, participants view images that fly by them rapidly on a computer screen. Their goal is to make decisions about two targets that appear in the stream of information. The stimuli are presented so fast that they can barely consciously recognize them, but their brain's processing speeds are fast enough to keep up unless the two targets are close to one another in time (within a half a second). Under these circumstances, the participants are partially blind to the second target. Note that it is not because they blinked their eyes, but because it takes time for attention to be allocated again so soon after it was just deployed. It is as if their brain had blinked and needed time to turn on again. In our scenario, if a bird exploded into sight from the brush it might have caused an attentional blink for our ancestor, preventing him from identifying the jaguar at that key moment in time. It is easy to imagine how such limitations on processing speed affect us when it comes to the fast-paced world of highway driving.

Another aspect of limitations on the speed of attentional processing is that it not only takes time to allocate our attention when we want to, but it also takes time to disengage our attention if it was captured by bottom-up influences.[14] Imagine a situation where your attention is captured by an irrelevant piece of information, such as hearing your name mentioned at the next table at a restaurant. Even though you already realized it was not you being referred to, it still captures your attention against your will because of the strong bottom-up salience of your name (a limitation in your selectivity). But then it also takes time for you to withdraw your attention from this distraction, and to reallocate it to the conversation you were having at your table. Even for a simple distracting stimulus, the "recovery time" from having your attention captured takes tenths of a second. And so, the speed of attention, both its allocation and disengagement, represents yet another limitation of this cognitive control ability.

Just as selective attention has these inherent limitations, working memory has them as well. These have been characterized in two domains: *capacity* and *fidelity*.[15] Capacity refers to the amount of information being stored, often in terms of the number of items that can be held in mind at any given time. Fidelity refers to the quality, or resolution, of the stored internal representations of those items—how faithful they are to what they represent. Encapsulated in the construct of fidelity is the rate of decay of a stored representation over time. All memory systems, including our computerized ones, can be described by these two characteristics.

Working memory capacity has been a major area of focus in the field of cognitive science, especially when it comes to research directed at understanding the limitations of our cognition. One of the most famous papers in the field was published in 1956 by the psychologist George Miller with the title "The Magical Number Seven, Plus or Minus Two: Some Limits on Our Capacity for Processing Information."[16] In this paper, Miller described our limited capacity to store information as having a span that is often defined simply as the longest number of items that can be immediately repeated back in the correct order. You can try this for yourself. Have a friend write down a list of single-digit numbers, and read them to you one at a time and ask you to repeat them back in order. In all likelihood you will be able to recall between five (seven minus two) and nine (seven plus two) items. Psychologist Nelson Cowan concluded that when studies are performed that prevent rehearsal and "chunking" of information (such as putting together three consecutive numbers that happen to be your area code), our true capacity limit is more like four, plus or minus one.[17] Other studies have shown that the type of information also influences the span of our working memory.[18] Thus, capacity may be seven for digits, but six for letters, five for words, only three or four for objects—and for stimuli as complex as a face, the working memory span may be two or even one. The point is that is our working memory capacity is quite limited. Individual differences in capacity have been shown to be associated with higher-order cognitive abilities related to real-world activities, such as reading comprehension, learning, and reasoning, as well as estimates of intelligence.[19] Individuals who have a larger working memory capacity tend to do better on assessments of these skills and measures of general fluid intelligence.

In addition to capacity, the other notable limitation of working memory is the reduced fidelity of information that is stored in mind, as well as the rapid rate of decay in information quality over time. We can appreciate based on our experiences alone that our internal representations of information do not have the same level of detail that existed when those stimuli were first in front of us (i.e., as in perception). This has been demonstrated formally in experiments showing that the transition from perception to working memory involves a loss of precision of detail.[20] Take a look at a crowded room and let your eyes roam across the people who are standing around talking. When you close your eyes, do you see an exact snapshot of the room? What about after your eyes have been closed for ten seconds? Information in working memory decays rapidly.

There is still debate over the exact source of this decay in our working memory: does it occur merely as a result of the passage of time or is it caused by interference? It seems that both of these are likely occurring. The act of holding information in mind is an active process, similar to how resources are required to sustain attention to an external stimulus. And so, even without interference, our memories decay because of fluctuations in our ability to sustain working memory over time. In addition, it is now clear that our ability to hold information in mind is fragile and susceptible to interference not only by interruption, but also by distraction. In an experiment similar to the Gazzaley Lab face/scene experiment described in the previous chapter, we instructed our research participants to hold the image of a face in mind for seven seconds, after which we tested their memory of that face.[21] The twist here was that on some trials we flashed a picture of a different face on the screen halfway through the delay period—when they were holding the original face in mind. They were told beforehand that this would happen and that this face was entirely irrelevant and should be ignored. Despite the warning, their memory for the single face was subtly, but consistently, reduced when a distracting face popped up. This shows us that even healthy young adults are easily distractible while performing a very simple working memory task. We further revealed that those participants who showed increased visual brain activity for the distracting faces performed more poorly on the working memory test.[22] These results demonstrate that overprocessing irrelevant information diminishes the fidelity of a working memory trace. Cognitive scientist Dr. Edward Vogel and colleagues showed that distraction also influences working memory capacity, in that individuals who exhibit greater distractibility by irrelevant information have a lower capacity.[23]

Our ability to maintain information in mind is very vulnerable to interference by both distraction and interruption. Try to recall how difficult it is to hold detailed directions of where you are going in your mind while dealing with the high-interference scenarios of managing traffic. Throw in the radio blaring and a text message arrival and your directions are out the window.

GOAL-MANAGEMENT LIMITATIONS

We make a choice between two options when we decide to accomplish more than one goal within a limited time period: to multitask or task switch. Sometimes we decide to try to do two things at exactly the same time, such as talking on the phone while reading email (multitasking), and sometimes we decide to switch between tasks, such as writing a paper and then flipping over to read an incoming email (task switching). Multitasking and task switching are distinct behaviors, but their enactment in the brain proceeds largely via the same mechanism: network switching.

It is more obvious that we are switching between information-processing streams when we are writing a paper and responding to email, or driving in a car and texting, because they use the same sensory system; we actually need to look away from one source and direct our vision to the other. But this is essentially what happens when attempting to talk on the phone and respond to an email or drive and talk on the phone hands free. We rapidly switch back and forth between these tasks, even if it was not our intention to do so, and even if we are not aware that we are doing it. If you really try to pay attention to how you multitask with this in mind, you will likely be able to tell that switching is occurring. Try listening to a television reporter while reading online. When you are reading, can you understand what the reporter is saying? Many of you have had the experience of trying to check your email while you have a conversation on the phone. At some point you are likely to lose the thread of conversation, and have to reengage without making it obvious that you were not paying full attention.

Our brains do not parallel process information, as demanded by many of our daily activities, if those activities both require cognitive control. This failure of our brain to truly multitask at a neural level represents a major limitation in our ability to manage our goals. The process of neural network switching is associated with a decrease in accuracy, often for both tasks, and a time delay compared to doing one task at a time. Known as multitasking or task-switching *costs*, these decrements in performance occur for both types of goal management. You can think of these costs as the price you pay for trying to do more than one thing at a time.

Interestingly, the term "multitasking" did not originate in the field of psychology or neuroscience, but was borrowed from the computer world where it refers to parallel processing of jobs.[24] Although some powerful computers do indeed simultaneously process multiple operations and thus engage in true multitasking, not every computer is capable of this. Single-processor computers—including our tablets and smartphones—end up doing something much more akin to what our brains do when we ask them to perform multiple operations at the same time. Apple shared an interesting description of this very phenomenon in commercials when the iPhone operating system 4.0 was first released. They were proud to announce the launch of a new feature that was in high demand: "multitasking." Apple claimed that their new iOS operating system used an approach to multitasking that was designed to save battery life. How did they pull this off?

> The reason is simple: This is not 100% true multitasking. All system resources are available to all applications, with the system assuming the role of a traffic controller, giving preference to some tasks and less preference to others as needed.[25]

This is essentially how our brains manage multiple tasks that require cognitive control: the prefrontal cortex serves as a traffic controller to facilitate goal management by rapidly switching between neural networks associated with each of the tasks. It is fascinating to read Apple's description for why they went this route, rather than implementing true multitasking in their phones:

> Free-for-all multitasking will consume way too many resources, especially memory. This will make the system choke, given the limited memory available in these devices. The CPU would also be taxed, and it would deplete the battery life quicker while slowing down applications running on the foreground.[26]

Such a description could easily have been written about our brains, rather than our iPhones. Perhaps a reason why our brains did not evolve true multitasking capabilities is that the competition for resources involved in cognitive control would also choke the system and create an energy drain.

The Gazzaley Lab has been assessing different types of multitasking costs and figuring out what happens in the brain when we try to multitask so that we can better understand this fundamental limitation to our cognitive control. We have shown that performing a secondary task that demands selective attention while simultaneously holding information in mind diminishes working memory performance. This occurs both when the information held in mind is something complex, like a face, or something very simple, like the direction of a moving field of dots.[27] In both cases, brain activity recordings revealed that the more an interrupting task was processed in the visual cortex, the worse the performance was on the ongoing working memory task. Also, those participants who struggled to reengage the prefrontal cortex network associated with the working memory task after they were interrupted performed worse on the working memory task. These studies revealed that the neural mechanisms that result in multitasking costs were both processing of the secondary task and a failure to effectively switch back to the original network after being interrupted.

We also studied another type of multitasking that does not involve working memory, but rather competition between two selective attention tasks in a video game.[28] In this study, we instructed our participants to make rapid and accurate responses only to target signs (e.g., a green sign) and ignore distractor signs (e.g., a red sign), thus demanding selective attention. On some trials they had do this task while simultaneously driving a virtual car in a 3D environment, which involved a great deal of selective attention to maintain position on the road. We found that twenty-year olds, despite great confidence in their multitasking abilities, suffer a significant multitasking cost: accuracy on the sign task dropped by 27 percent when they navigated the car at the same time. Thus, goal enactment that requires cognitive control, whether it is the domain of working memory or selective attention, is detrimentally impacted by a secondary goal that also demands cognitive control.

However, you do not have to be multitasking, or attempting to simultaneously complete two tasks, to exhibit performance costs. We experience costs even when we explicitly decide to switch between tasks. This is true even for simple tasks, as long as they demand some degree of cognitive control. You can observe this for yourself. Try the following exercise:

First count from one to ten out loud. Then recite the alphabet from A to J out loud. Those two tasks should be quite easy. But now try combining them by rapidly switching between them: recite A1, B2, C3, and so forth out loud. You will likely sense the limitations imposed by the network switching that is required to do this, as well as notice the cost of being slower—perhaps even making an error or two if you push yourself to perform this rapidly.

Task-switching costs are often assessed in the lab as the time difference between how long it takes to complete a task immediately after performing an identical task, compared to the time it takes to perform that same task immediately after a different task. Task-switching costs calculated in such a manner reveal that there is a cost even if you can predict when a switch will occur. Although costs do occur even for simple tasks, they certainly increase with task complexity.[29] And so, limitations in goal management exist both when we attempt to multitask and when we try to switch between tasks.

SUMMARY OF COGNITIVE CONTROL LIMITATIONS

We have described many limitations that exist in our cognitive control abilities across the domains of attention, working memory and goal management. Here is a summary of the limitations described throughout the chapter.

Attention

1. *Selectivity* is limited by susceptibility to bottom-up influences.
2. *Distribution* of attention results in diminished performance compared to focused attention.
3. *Sustainability* of attention over time is limited, especially in extended, boring situations.
4. *Processing speed* limitations affect both the efficiency of allocation and withdrawal of attention.

Working Memory

1. *Capacity*, or the number of items that can be actively held in mind, is severely limited.
2. *Fidelity*, the quality of information maintained in working memory, decays over time and as a result of interference.

Goal Management

1. *Multitasking* is limited by our inability to effectively parallel process two attention-demanding tasks.
2. *Task switching* results in costs to accuracy and speed performance.

Understanding the limitations in our cognitive control is critical to appreciate the conflict between our goals and our limitations that are the core thesis of this book. In the next chapter, we discuss how the Distracted Mind is not a constant entity; rather, it varies across individuals and exists in a state of flux.

5 VARIATIONS AND FLUCTUATIONS

IF WE ARE to develop effective approaches to alleviate the Distracted Mind, we must first understand that it is not an unmovable, static entity. Our cognitive control abilities, and the flip side, our cognitive control limitations, are not completely fixed traits that define us. Rather, they exist in a state of flux, morphing over time, from day to day and over the course of our lives, and as a result they come under the influence of numerous factors. Variations and fluctuations are the rule, not the exception.

One of the most studied aspects of the shifting Distracted Mind is how our cognitive control abilities change throughout our lives. They are at their worst in young children, where control limitations derail much of the goal-directed activity that starts to emerge from their developing brains. Cognitive control gradually and steadily matures and develops throughout the course of brain development, reaching a peak in the early twenties. Of course, even at its height in young adulthood it is still capped by those intrinsic limitations described in the last chapter. This pinnacle of cognitive control is immediately followed by a slippery slope of decline as we approach middle age, with an almost ubiquitous march downward as we advance into our senior years.

Cognitive control limitations are influenced not just by the passage of time, but also by the many pathologies associated with clinical conditions that degrade brain functioning at different times of life. Five common conditions that we will discuss here—ADHD, PTSD, traumatic brain injury

(TBI), depression, and dementia—are all associated with diminished cognitive control abilities that exacerbate the Distracted Mind.

Even for a healthy brain at any age, cognitive control abilities fluctuate from day to day, and even over the course of a single day. This is the result of many powerful influences such as sleep deprivation, psychological stress, and alcohol intoxication. Understanding how our control abilities are affected by these factors is critical to being able to "take control of control," as we will discuss in the final two chapters. Let's begin by reviewing what we know about age-related changes in cognitive control, stretching all the way from childhood to our senior years.

AGE-RELATED CHANGES

The Young Ones

Our cognitive control abilities improve gradually over the course of early childhood and reach peak levels in young adulthood (early twenties). This general finding is consistent for all control abilities—attention, working memory, and goal management—although the developmental patterns of different subcomponent processes vary, supporting the notion that cognitive control is not a singular construct.[1] As expected, this time course of improving cognitive control abilities across childhood maps directly onto the functional development of the neural mechanisms of cognitive control—that is, the development of top-down modulation—and parallels the course of structural brain development of the prefrontal cortex and its networks with the rest of the brain.[2] This protracted maturation is rather specific for the prefrontal cortex, and is not true of all brain regions—notably, it is not true of the motor cortex and the sensory cortex, which mature much earlier in life.[3]

When it comes to attention, we have extensive evidence of this protracted developmental trajectory. Selectivity, as assessed by a child's ability both to ignore irrelevant information and to search through a scene for relevant information, gradually improves from childhood through young adulthood.[4] Before top-down modulation is fully developed, the strong bottom-up influences that compete for our attention have much more impact on young minds. Most of us have witnessed a child easily distracted by a shinier toy, even after she was completely enthralled by the toy that she was playing with at the time. This phenomenon persists well into the teenage years, as the prefrontal cortex continues its slow development and results in poor decisions that extend beyond toys, much to the frustration of parents everywhere. This is not just limited to the selective aspect of attention, but also includes protracted development of a child's ability to distribute their attention, challenging performance in acts such as driving.[5]

Researchers have described a similar developmental pattern for the ability to sustain attention over time.[6] There is likely not a single teacher who does not struggle with this factor in the classroom, especially with younger children. The act of staying seated throughout the day is quite difficult for most children, and the dry nature of the content further compounds the situation. Many teachers are driven to develop clever ways of combating children's poorly developed abilities to sustain attention, such as using problem-solving exercises to increase interactivity.[7] This challenge in the classroom is likely aggravated by limitations in children's working memory capacity, which is also quite slow to develop. Young children can hold only a few items in mind; this capacity gradually increases throughout adolescence, but even more slowly for complex information.[8]

And then there are goal-management abilities. Many of us expect (or worse, demand) adult-level, goal-enactment abilities from children and young adults. This often generates extreme frustration and strained relationships when they act in ways that seem illogical, such as beginning tasks and then rapidly abandoning them as they unsuccessfully attempt to multitask or switch between tasks. And even when children gain the skills necessary to engage successfully in higher-level goal management, studies show that their performance costs are more pronounced than those of adults, especially on more demanding tasks.[9] It is important to recognize also that these immature cognitive control abilities are being used to attempt to navigate a high-tech ecosystem that was designed explicitly to encourage multitasking and task switching. Dr. Rosen's research addresses this directly, and will be covered in detail in Part II, where we will focus on real-world behaviors of the Distracted Mind, largely in the context of modern technology.

The Senior Years

Let's leap ahead and discuss what happens to cognitive control at the other end of the life span. Studies that compare the performance of sixty- to seventy-year olds with twenty- to thirty-year-olds on tasks that demand cognitive control are the most common research approach for exploring the effects of aging. Although there are numerous caveats and complexities in drawing conclusions from this type of cross-sectional research, the weight of the evidence supports the conclusion that our cognitive control abilities diminish as we age, and this leads to poorer performance on a wide range of goal-related tasks and activities.[10] This is believed to be largely independent of the many pathologies of aging that result in dementia, such as Alzheimer's disease, as indicated by brain data showing that the prefrontal cortex is one of the earliest areas to show signs of age-related degradation associated with the healthy aging processes.[11] In this section, we will share evidence from the Gazzaley Lab and other labs showing diminished abilities in all aspects of cognitive control in older adults, as well as the underlying neural mechanisms of these age-related cognitive changes.

When it comes to attention, there is extensive evidence that older adults have greater limitations than younger adults across all domains: selectivity, sustainability, distribution, and speed.[12] In terms of selectivity, age-related deficits in attention have been well documented for attention selectively directed at features, objects, locations, or moments in time. The Gazzaley Lab has been very interested in understanding the neural basis of this attentional selectivity deficit. In one study that used the face/scene experiment described in chapter 3, we challenged the selective attention of older research participants by having them view a sequence of four images (two faces and two scenes) and asking them to remember the stimuli from only one category for a brief period of time while ignoring the stimuli from the other category (e.g., remember faces and ignore scenes). As described, young adults show both enhancement and suppression of neural activity in the visual cortex in response to the relevant and irrelevant images, respectively. This bidirectional nature of top-down modulation of neural activity is at the core of attentional selectivity. Our main finding in this study was that, interestingly, older adults enhance activity for relevant information as well as twenty-year-olds. Where older adults suffered a deficit was in suppressing the irrelevant information. Thus, we discovered that their main attentional issue was that they are more distractible than younger adults.[13]

The Gazzaley Lab published the results of this study as evidence that selectivity impairments in older adults were the result of a neural deficit in the mechanisms responsible for goal-directed, top-down suppression. Even though our older research participants were aware of the instructions to ignore the irrelevant information, we did not see evidence in their brains that they were effectively suppressing the processing of those stimuli. We went on to show the functional significance of this deficit by demonstrating that those older adults who exhibited poorer suppression of distractors also did more poorly on the working memory test.[14] Note that this reduced performance is occurring in the context of a preserved ability to focus on relevant information. This is further evidence that the processes that underlie focusing and ignoring are not two sides of the same coin. Older adults have impairments in suppression, but not enhancement, and this is the deficit that is related to their diminished working memory performance. Further studies by the Gazzaley Lab have generalized this result by showing that the same findings exist for selective attention to visual features and moments in time, although other labs have shown mixed results for attention to spatial locations. Suppression deficits are even evident when older adults are given more time to prepare for an upcoming distractor.[15] There is a growing consensus of research studies that support the conclusion that attentional deficiencies in selectivity are not the consequence of an inability of older adults to focus on their goals, but rather are the consequence of a selective deficit in *ignoring distractions.*

We have recently shown that this selectivity deficit is associated with age-related alterations in prefrontal cortex networks; and not just functional changes, but also structural changes in the volume of a region in the middle part of the prefrontal cortex, as well as diminished integrity of the white matter that connect this area with other brain structures. We also found that older adults with these brain changes were more distractible on a working memory test.[16]

Another study from the Gazzaley Lab showed that this attentional selectivity deficit in older adults was related to another limitation in attention abilities discussed in the last chapter: the speed of attention processes. Using EEG recordings, we found that it is not that older adults completely lack the ability to suppress processing of distractions, it is just that they do not do it rapidly enough.[17] Impressively, young adults suppress irrelevant information within one-tenth of a second after they confront a visual distraction. Older adults, however, do not show neural signs of suppressing the brain activity associated with a distracting face until at least half a second. These results suggest that if distractions are not suppressed almost immediately, they have time to create interference with the processing of relevant information, in turn degrading both working memory and long-term memory performance.[18] In other words, our distraction filter needs to stop the flow of noise from entering our brain at the entrance gate. If you are an older adult sitting in a noisy restaurant trying to engage in conversation and you are not rapidly filtering out the chatter, it is going to make following the conversation challenging. This is likely the source of anecdotal reports we hear from older adults who do not enjoy the restaurant experience as much as they once did. Unfortunately, selectivity deficits along with accompanying delays in processing speed are not even the whole story; limitations in sustainability and distribution of attention also contribute to attention challenges experienced by older adults.

Working memory shows similar age-related declines. Research from the Gazzaley Lab has shown that healthy older adults exhibit diminished fidelity of working memory, even for holding a single item in mind for several seconds, whether it is a face or a simple visual feature.[19] Other labs have shown that it is not just fidelity that declines, but the capacity of working memory also diminishes with age, although there is debate over whether this is a pure working memory issue or a consequence of older adults being more distractible and slower in general.[20] Related to working memory abilities, older individuals also exhibit diminished mental imagery, which is the ability to re-create a scene in mind from long-term memory.[21] We showed that this was associated with diminished selective activation of prefrontal cortex networks, echoing a repetitive theme that age-related deficits in cognitive control stem from dysfunction in the prefrontal cortex.[22]

When it comes to goal management, older adults once again have more difficulties than young adults; this is true of their ability to engage in more than one task at a time whether it be multitasking or task switching.[23] The Gazzaley Lab performed a series of brain imaging studies to understand exactly what occurs in the brains of older adults that underlies their greater performance impairments when they attempt to simultaneously engage in two cognitively demanding tasks.[24] We had older adults perform an experiment where we instructed them to remember faces or scenes for a short period of time, and then interrupted them with another task while they were holding the relevant information in mind. We found that their working memory performance was impaired more by these interruptions than it was for young adults, and this was associated with less effective switching between the prefrontal cortex network involved in holding the information in memory and the prefrontal cortex network engaged in performing the interrupting task. Other labs have reported similar network deficits in older adults engaged in task switching and multitasking challenges.[25]

Interestingly, the brain changes that underlie goal interference deficits experienced by older adults—distraction and interruption—are mechanistically distinct: distractibility is caused by an inefficient filter that results in excessive processing of irrelevant information in the visual cortex, whereas multitasking impairments are caused by a failure to effectively switch between networks involved in performing two tasks. In common between them are alterations in prefrontal cortical networks, both functional and structural, which are at the heart of their goal-interference issues.

In summary, a large body of scientific research has revealed evidence of age-related impairments, independent of dementia, in all cognitive control abilities, which is associated with worsened goal interference as we age. Interestingly, these deficits do not seem to be accompanied by less lofty goal-setting behavior. Older adults by all indications are in the process of reinventing what it means to be "older." It is increasingly common for older adults to remain engaged in the workplace until later in life, and when they do retire, they often pursue high-level challenges, such as traveling, learning foreign languages, and mastering musical instruments.[26] They also do not seem to shy away from the high-interference behaviors offered by the high-tech world. This conflict between increasingly more ambitious goals and greater limitations in cognitive control is at the core of the older Distracted Mind.

ACROSS THE LIFE SPAN

We have already described differences in cognitive control abilities as a function of age and discussed how they peak over the course of development and decline once we are over sixty years old. This is consistent with the role of the prefrontal cortex in cognitive control and how this essential brain structure changes over the years; it is the last part of our brain to develop and the first to degrade in our senior years. But what happens along the full path of our lives? Do we maintain our control abilities at peak levels throughout our adult years, and only after many decades of living the good life do they dramatically plunge? As it turns out, this is not the case. In general, cognitive control abilities exhibit a steady decline across the adult life span.

Let's take a look at the pattern of change in working memory capacity across several different tests from ages twenty to ninety (figure 5.1). That downward trajectory, which is quite linear, is a common pattern that is observed for the full array of cognitive control abilities. Although the trajectory varies for different subcomponents of these abilities, this likely reflects distinct patterns of change across subregions of the prefrontal cortex. As a side note, although this pattern of decline is found for cognitive control, this does not reflect how *all* cognitive abilities change with age. Some aspects of cognition, such as vocabulary, remain relatively constant (perhaps even increasing) throughout our adult lives.[27]

To get the complete picture of life-span changes, several research teams have gone so far as to examine cognitive control all the way from childhood to older age. These studies reveal a fascinating U-shaped pattern. Figure 5.2 shows two examples of curves (note: lower values reflect better performance since this is a response time measurement), one for an attention task and another for task switching.[28] While this pattern of improvement until the late teens or early twenties that is followed by a linear decline in abilities with each passing decade has been consistently shown for cognitive control

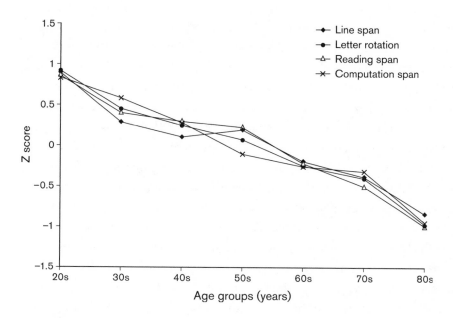

Figure 5.1

The downward slope of working memory capacity across the adult lifespan. From D. Payer and D. Park, "Working Memory across the Adult Lifespan," in *Lifespan Cognition: Mechanisms of Change*, edited by Ellen Bialystok and Fergus I. M. Craik (2006), 131, Part A, figure 9.1. By permission of Oxford University Press.

abilities across many studies, it is important not to oversimplify this and assume that it means that aging is simply the reverse of development. There are actually different mechanisms that account for developing these skills versus their decline.

This life-span perspective of cognitive control is critical for appreciating one of the most salient aspects of the Distracted Mind: it varies dramatically across the life span. The Gazzaley Lab further explored this lifespan phenomenon in a series of experiments that used a video game the lab developed known as *NeuroRacer* (discussed further in chapter 10). Playing this game is fairly straightforward. You sit in front of a laptop with a joystick in hand while signs of different colors and shapes pop up on the screen in front of you every few seconds. You are instructed beforehand which of the signs is your target (e.g., a green circle) and told that when it appears you need to press the button on the joystick as rapidly as possible, while avoiding pressing the button for any distracting signs (e.g., a red circle or a green pentagon). You are scored on how accurately and rapidly you perform this task, known as the "single-task version." Things get more interesting in two other versions of the game. In the "distraction version," the goal is exactly the same, except now there is a colorful 3D road stretching out into the distance, and there is a car in front of you that is driving along that road. However, you don't have to actually drive the car; it is on autopilot. Your goal is the same as in the single-task version: to direct your attention to respond only to target signs and to ignore the distractions of the road and moving car. The final version of the game, known as the "multitasking version," has you manage two tasks simultaneously. You have to perform the sign task, in the same way you would in the single-task and distraction versions, but now you also have to navigate the car along the road. This requires that you to move the joystick left and right when the road turns, and push forward and backward as you move up and down hills in order to maintain a constant speed. Navigating the car requires a great deal of attention to avoid going off the road or crashing into speed markers in front of and behind your car; and all the while, those signs keep popping up.

Research participants engaged in all three versions of *NeuroRacer* over a single day of game play in the Gazzaley Lab. The data allowed us to determine how susceptible each person was to distraction by comparing performance on the single-task and distraction versions, and how well they multitask by comparing performance on the single-task and multitasking versions. Interestingly, we found similar patterns across the life span for both types of goal interference. In terms of distraction, even though there was only one goal, the mere presence of the moving road distracted the participants and resulted in diminished performance on the sign task. This distraction effect was present in eight- to twelve-year olds, almost nonexistent for those in their twenties, and then gradually worsened with each passing decade.[29] In terms of multitasking, the impairment in the sign task induced by driving was even more dramatic than it was for distraction. But once again, the same pattern emerged (figure 5.3), supporting the takeaway message that

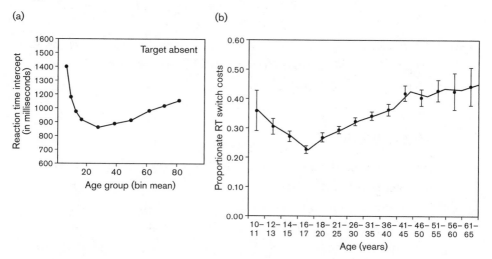

Figure 5.2

Attention and task switching performance reveal a U-shaped pattern from childhood to our senior years, with abilities improving until young adulthood (depicted as a lower value) and then declining with advancing age. Part A adapted from B. Hommel, K. Z. H. Li, and S.-C. Li, "Visual Search across the Life Span," *Developmental Psychology* 40, no. 4 (2004): 545–558 (fig. 3). Part B adapted from S. Reimers and E. A. Maylor, "Task Switching across the Life Span: Effects of Age on General and Specific Switch Costs," *Developmental Psychology* 41, no. 4 (2005): 661–671 (fig. 5).

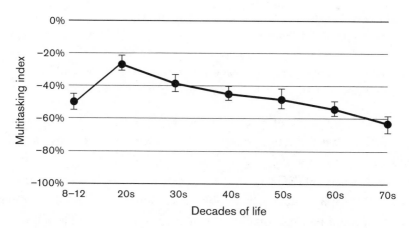

Figure 5.3

Multitasking abilities exhibit a pattern of improvement until early twenties followed by a linear decline into our senior years. Unpublished data from the Gazzaley Lab.

the Distracted Mind is not a constant across our lives, but varies in terms of its susceptibility to the negative effects of goal interference, both distraction and interruption, on task performance.[30]

STATE CHANGES

Susceptibility to interference varies not just over the course of our lifetime but also fluctuates based on our current state. Let's begin by considering the difference between traits and states. Traits are a relatively static reflection of an individual that do not vary much over time—certainly not from day to day. They are the type of factors that you might use to identify someone, for example, eye color. Personality features of an individual are less fixed but remain relatively constant from day to day, and so are sometimes also considered traits; kindness, for example, may be a considered a personality trait if it is consistently displayed. States, on the other hand, are truly in flux, and do not define an individual; for example, someone may be feeling tired one particular morning. More often than not, individuals exhibit a complex mixture of states and traits; while there may be overall consistency that defines a trait, there may also be susceptibility to state influences. Here is an ancient description of states and traits that captures their essence:

> It is one thing to be irascible, quite another thing to be angry, just as an anxious temper is different from feeling anxiety. Not all men who are sometimes anxious are of an anxious temperament, nor are those who have an anxious temperament always feeling anxious. In the same way there is a difference between intoxication and habitual drunkenness. (Cicero, 45 BC)[31]

This mix of state and trait is certainly the case for cognitive control abilities. A surprising result of a twin study revealed that a self-report distractibility measure showed a significant degree of genetic influence, suggesting that at least some aspects of the Distracted Mind may indeed be inherited as traits.[32] And so, a given individual may exhibit distractibility as a trait, if it is a consistent aspect of their personality, but their level of focus will also fluctuate from day to day as influenced by various state effects. Sleep deprivation, psychological stress, and alcohol intoxication are three common factors that induce powerful influences on cognitive control, and thus impact the Distracted Mind. In general, their presence tends to degrade cognitive control, with more extreme and longer-lasting effects occurring with chronic exposure. This has been assessed using both laboratory assessments and real-life activities, such as driving.

As you may have surmised from observations of your own mental sluggishness after a bad night's sleep, acute sleep deprivation negatively impacts cognitive control in a major way. Notably, it has been shown to impair sustained attention.[33] This is exacerbated by a tendency to induce bouts of microsleep—brief periods of time when the brain slips rapidly into sleep mode. However, as you may have also observed, individuals show substantial differences in how susceptible they are to the effects of sleep deprivation.[34]

Researchers are expanding their work on sleep disturbances, particularly focusing on how lack of sleep or poor sleep affects brain functioning during the daytime. One study, for example, took brain scans of adults three and a half years apart and found that those who had the most sleep difficulties showed a more rapid decline in brain volume.[35] More specifically related to cognitive control, however, is research showing that just one night of poor sleep can lead to less efficient filtering out of important information from junk as well as inefficient visual tracking, both of which, of course, underlie the Distracted Mind.[36] In addition, other researchers showed that adolescents with poorer sleep showed less recruitment of the prefrontal cortex during cognitive tasks in addition to reduced network communication between the prefrontal cortex and brain regions that process rewarding experiences, which is then linked to more risky behaviors.[37]

Disturbed sleep has also been linked to other cognitive issues that impair our ability to maintain focus. For example, one study of seven- to eleven-year-old children in Quebec, Canada, asked parents of one group to have their children go to bed earlier than normal (averaging a bit more than half an hour of additional nightly sleep) while the other half went to bed an hour later than normal. Classroom teachers rated their behavior without knowing which group they were in and found that the sleep-deprived children showed reduced cognitive control, particularly in the areas of attention, increased impulsivity, and frustration.[38] As Dr. Judith Owens, director of sleep medicine at Children's National Medical Center in Washington, DC, explains, "We know that sleep deprivation can affect memory, creativity, verbal creativity and even things like judgment and motivation and being (engaged) in the classroom. When you're sleepy, [being engaged] isn't going to happen."[39]

When it comes to the impact of stress on cognitive control, the story is a bit more complex. The study of psychological stressors has been complicated by difficulties in sorting through the influences of source, duration, intensity, timing, and challenge on stress. These factors have a big impact. For example, researchers have established that even the direction of influence is affected by the intensity of stress: some amount of stress can be beneficial, while excessive amounts are harmful to performance. This relationship is further modified by task difficulty. Consider the Yerkes–Dodson law, an inverted-U pattern described by psychologists Robert Yerkes and John Dodson at the beginning of the twentieth century. It describes the relationship between arousal and performance, such that performance increases as arousal (or stress) goes up, but then once a certain level is reached performance begins to decline—that is, if the task is sufficiently difficult (figure 5.4).[40] The complicated relationship between stress and cognitive control is emphasized by the presence of studies that show stress impairs working memory and attention and others studies that propose stress improves these abilities.[41]

It likely should come as no surprise that alcohol intoxication impairs cognitive control. It's not subtle. Numerous studies have documented its negative impact on tests of working memory, selective attention, sustained attention, and multitasking, which actually occur at relatively low levels of alcohol exposure (<.05 percent, with 0.08 percent defining "drunk driving" in most states).[42] Of further interest is that, during the recovery period of impairment from acute alcohol intoxication, speed of performance on activities that require cognitive control (such as driving) recovers faster than performance errors, which may actually increase as blood alcohol levels diminish. Thus, we tend to be able to perform more quickly during this time, but we make more mistakes. This is certainly something to keep in mind if you are thinking that you have recovered enough from a night of drinking to get safely behind the wheel.

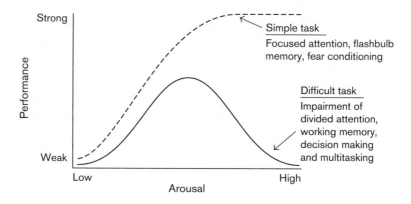

Figure 5.4

Yerkes–Dodson curve based on the original evidence from Yerkes and Dodson (1908). From D. M. Diamond et al., "The Temporal Dynamics Model of Emotional Memory Processing: A Synthesis on the Neurobiological Basis of Stress-Induced Amnesia, Flashbulb and Traumatic Memories, and the Yerkes–Dodson Law," *Neural Plasticity* 33 (2007), doi:10.1155/2007/60803. PMID 17641736. © 2007 David M. Diamond et al. CC BY 3.0.

CLINICAL CONDITIONS

As described, our cognitive control abilities emerge from complex neural network interactions that require finely tuned and rapid communication between brain regions. Thus, pretty much *anything* that negatively affects the functioning of our brain will erode the quality of cognitive control abilities. This then creates more conflict with accomplishing our goals, generates more interference, and results in a more Distracted Mind. This is the reality for most of the psychiatric and neurological conditions that you are familiar with: ADHD, PTSD, TBI, major depression, schizophrenia, and Alzheimer's disease. For all these conditions, deficits in cognitive control exacerbate the other clinical symptoms and often cause the most negative impact on quality of life.

Despite the overlap in cognitive control deficits, these clinical conditions present in very different ways and at different times of life. They may develop over the course of childhood and adolescence (e.g., ADHD, schizophrenia) or emerge as degeneration at an older age (e.g., Alzheimer's disease). Others may be equally likely at any age, and relatively independent of life events (e.g., major depression), or they may be situational and highly dependent on life events (e.g., TBI and PTSD). Let's touch on each of the components of cognitive control and consider how the presence of these clinical conditions in an individual aggravates an already Distracted Mind.

When it comes to attention, it should come as no surprise that the clinical condition that earned the name in its classification—attention deficit hyperactivity disorder—is associated with major deficits in this domain. Starting with selectivity, children and adults who are diagnosed with ADHD are more susceptible to be negatively influenced by distractors than their age-matched peers.[43] But those diagnosed with ADHD are not alone in experiencing this burden. On tests of distractibility, all the clinical conditions mentioned in the preceding paragraph exhibit marked impairments. Consider the Stroop test, a classical assessment of distraction where you see words that are written in colored ink and have to say out loud the name of the color, not the name of the written word. For example, if you see the word RED written in blue ink, you should say "blue" rather than "red." When the color ink and the name are matched (e.g., the word RED in red ink) the response is much more rapid than when RED is written in blue ink. Results of the Stroop test reveal difficulties in suppressing task-irrelevant information for individuals with ADHD, traumatic brain injury, depression, post-traumatic stress disorder, schizophrenia, and Alzheimer's disease.[44]

The attention deficits experienced by individuals with these conditions extend well beyond the domain of selectivity, notably when it comes to the act of sustaining attention. Deficits in sustained attention are one of the most consistently reported issues in children with ADHD, which involves underactivation of the prefrontal cortex.[45] Impaired sustained attention abilities have also been documented in individuals with PTSD, such as rape survivors and war veterans, who still display deficits years after the traumatic event.[46] Similar findings have been reported for the coexistent conditions of TBI and major depression.[47] In Alzheimer's disease, attention deficits are the first non-memory-related impairments to appear, but while selectivity of attention is affected early in the disease, sustained attention deficits occur later in the course of the illness.[48]

Working memory performance, when tested both verbally and visually, is impaired in patients with ADHD, with some experts considering this deficit to be a core aspect of their functional impairment.[49] Evidence of deficiencies in working memory have also been reported in patients with PTSD, and this has been linked to lack of engagement of prefrontal cortex networks.[50] This aspect of cognitive control is especially vulnerable in patients with TBI and is considered central to their cognitive impairment profile.[51] Once again, brain-imaging data revealed notable alterations in activity of the prefrontal cortex.[52] And rounding out the list, patients with both major depressive disorder and Alzheimer's disease also exhibit working memory impairment.[53]

Although attention and working memory are routinely impacted by psychiatric and neurological disease, in many ways they pale in comparison to the more severe deficits of goal management abilities. A survey of the scientific literature reveals that across the many different laboratory tests used to assess goal management there are widespread impairments in all the clinical conditions we have been discussing, as well as many others. Patients diagnosed with PTSD, TBI, depression, ADHD, and Alzheimer's disease all have greater difficulty when attempting to engage in two tasks at the same time when compared to age-matched control populations.[54] This is true not only when individuals are assessed using computer lab tests, but also using tests that are closer to real-world activities. For example, walking and talking are two tasks we engage in at the same time daily, and most of us perform this low-level multitasking perfectly well. After all, these are largely automatic acts that do not generate very much competition for mental resources. And yet the gait of Alzheimer's disease patients is affected when they talk at the same time.[55] This is believed to contribute to the very serious risk of falling in this population. Even a comparatively young group of Alzheimer's patients exhibited similar difficulties in talking while walking.[56] Deficits assessed by walking while performing other tasks have also been shown for multiple sclerosis and Parkinson's disease.[57] Widespread multitasking deficits truly highlight the exquisite sensitivity of the Distracted Mind to all disorders of the brain.

Part I of this book has focused on understanding the Distracted Mind in terms of what is going on in our brains that has created our interference dilemma. We have shown how our brains, despite their highly evolved goal-setting abilities, are ancient in some fundamental ways: our information-seeking behavior is an extension of primitive food-foraging behaviors, and our limitations in cognitive control are comparable to that of many other animals. It is our high-level goal-setting abilities, coupled with our drive to seek information, that lead us to engage in interference-inducing behaviors that put pressure on those limitations in our cognitive control. This conflict results in goal interference, which in turn leads to a broad array of negative consequences that we experience in our daily lives. But cognitive abilities are not the only factors that influence real-world behavior. Behavior is contextual; it is a product of not just how we think, but also what is going on around us. Our environment interacts with our cognition in complex ways to yield our behavior. In Part II, we build on the foundation we have created to show how modern technology aggravates the Distracted Mind and affects the quality of our lives in unexpected ways.

Part II

BEHAVIOR IN A HIGH-TECH WORLD

OUR BRAINS, as amazing as they are, suffer from interference that generates the Distracted Mind. But our Distracted Mind does not live in a jar; it lives in the real world. Our daily behavior is influenced by more than just how effectively our brains process information—it is influenced by our environment. Decades of research in Dr. Rosen's laboratory and by other groups around the world have shown that one of the major factors that contributes to the Distracted Mind is the emergence of modern technology. We begin in chapter 6 by sharing research that shows how modern high-tech has led us to jump from task to task and how prevalent this behavior has become. In chapter 7 we discuss the impact of this technology-driven interference on multiple aspects of our lives, including education, safety, workplace, relationships, sleep, and health. Chapter 8 looks at this impact in various populations whose minds are already more distracted at baseline, such as children, older adults, and individuals with clinical conditions including ADHD, depression, and autism, and explores why they are more susceptible to interference. Finally, in chapter 9 we revisit the MVT model introduced in chapter 1 and explain the role that increased accessibility of information, anxiety, boredom, and reduced metacognition play in aggravating our Distracted Mind.

6 THE PSYCHOLOGY
OF TECHNOLOGY

IN 1970, prior to the technological revolution that resulted in the information age in which we now all reside, Alvin Toffler, described by some as "the world's most famous futurologist," penned his classic book *Future Shock*, which warned that we were entering an era of "too much change in too short a time."[1] Following up a decade later with *The Third Wave*, Toffler further described this as a process of succeeding waves of technological innovation, each of which begins, reaches a peak, and then starts to decline as the next wave performs the same process.[2] Toffler's first wave—spanning three thousand years—was labeled an agricultural or agrarian wave, where technologies were designed to aid farming and replace the older hunter–gatherer society. The second wave, spawned by the development of the steam engine and the industrial revolution, replaced agrarian technologies with those of factories and production systems and lasted three hundred years, one-tenth as long as the first wave.

Toffler wrote *The Third Wave* in 1980, well before the appearance of smartphones and the Internet, and even a few years prior to the introduction of the graphical user environment of icons and pictures—at least to the masses—by the Apple Macintosh. At the time he wrote *The Third Wave*, Toffler labeled his third wave the "computer wave" but also referred to it as the "global village," "information age," "space age," and "electronic era." Toffler predicted that this wave of new technology would most likely last about thirty years, to be replaced by another starting somewhere during the

1990s while the third wave began its downward trend near the turn of the century.

Using simple math, it is possible to project that the fourth wave might have been as short as three years, following the pattern of decreasing lengths of his previous three waves (3,000 → 300 → 30 → 3) with each successive wave one-tenth as long, producing waves of technology that last mere months or days. Obviously that is not possible; but since Toffler's computer wave, it seems we have been experiencing a wave that consists of a successive series of rapid, short "wavelets," each lasting perhaps three to five years at the most. These wavelets reflect the progression of the information age from a standalone computer with limited capabilities to our current day where nearly everyone carries a powerful computer in his or her pocket or purse.

We can identify the fourth wave as the "Information Age," which is composed of a series of five smaller wavelets that all involve some form of information dissemination and gathering, whether in its pure form, as you might find in an article, or as part of personal messages, transmitted through texts, emails, tweets, and the like. For simplicity's sake we can refer to these wavelets as 4.1, 4.2, 4.3, 4.4, and 4.5. We propose that wavelet 4.1, the first part of the Information Age, started when the Internet began to gain traction in the 1990s, and we reveled in our ability to access information that before was accessible only through libraries, dictionaries, or encyclopedias. Wavelet 4.2—which we have dubbed the beginning of the "communication era"—was highlighted by the universal adoption and use of electronic mail. With new communication modalities came increased accessibility for information foraging, as every message carried information. America Online—who coined the somewhat obnoxious announcement "You've Got Mail"—helped spawn wavelet 4.2 with its dedicated email system, which was quickly followed by a bevy of competitors who offered electronic connection at a monthly cost.

Wavelet 4.3 denoted a major change in that we no longer were tied to a single location such as our work desk to gain access to information; instead we entered a "mobile era," with the rising popularity of rather large portable computers, which then morphed into laptops, notebooks, PDAs, netbooks, and eventually, early cell phones. These devices allowed us to gather our information—and communicate with others, which is, after all, a key form of information dissemination and gathering. This could occur anywhere we happened to be, and it ushered in the birth of ubiquitous Wi-Fi to allow us to access the web from schools, libraries, and, of course, coffee shops.

The next stage of the Information Age, wavelet 4.4, included "social communication" or "virtual communities" and changed our one-to-one communication via email to one-to-many communication via social media. Social media resulted in an explosion of the amount of information we could—and for many, felt we needed to—forage from our many "friends" on our many social media sites. The current Information Age wavelet, number 4.5, introduced a major game changer, as the rather limited-function cell phone was replaced by the "smartphone," with all of the functions of a computer—making our ability to tap into information even more portable—plus a telephone, a music player, a video player, a camera, and nearly every possible source of information at the touch of an icon or a few simple taps on a glass screen. As we will describe later in this chapter, the smartphone has put a cap on the Information Age represented by the fourth wave in Toffler's nomenclature, as it now allows us complete access to any and all forms of information at any time of the day or night, which, as we will see, may not be particularly healthy for the Distracted Mind. Each of the Information Age wavelets that make up the fourth wave—the Internet, email, mobile access, social communication, and the smartphone—provide a new source of information technology and, as we will demonstrate, myriad new sources of goal interference.

Is there the hint of a sixth wavelet as part of the Information Age on the horizon, or are we perhaps moving to a completely new wave? We believe that we are just now starting to see an entirely new wave, as our technologies evolve from information wavelets to technologies that are adapting to our bodies and our biological functions. This wave is just starting to ramp up, and it appears that it will be a merging of the final wavelets of the Information Age with biological and medical science. We are moving toward the implementation of technologies that revolve around our brains and our bodies, as evidenced by the European Commission's Human Brain Project and President Obama's BRAIN Initiative as well as the rise in neuropsychological research using tools such as fMRI, EEG, and fNIR (functional near-infrared spectroscopy) to learn about how and why our brains react to stimuli and situations and then to enhance our abilities using other high-tech tools (as we will discuss in chapter 10). Add to these research tools wearables and diverse presentation and recording technologies such as augmented reality, virtual reality, motion capture, smartwatches, brain stimulators, implanted sensors, iris scanners, and even 3D printers that can make human organs—and it looks like we are entering a new biotech wave. Only time will tell.

The bottom line is that the rapid advancement of Toffler's waves and the Information Age wavelets that have arisen over the past decade or so—and seemingly have not begun their descent—have led us to a life of constantly morphing technologies, each of which fulfills a different niche or perhaps a perceived need. Most importantly they draw our attention and focus to deal with the ever-expanding explosion of that indispensable commodity known as *information*. In this chapter, we will explore how technology captures our attention and creates interference with our goals as we attempt to sift through the vast streams of information that are directed at us on a minute-by-minute basis.

Some consumer experts point to a benchmark that when 50 million people have used a product it is considered to have "penetrated" society.[3] When you consider technology products, this model made sense, at least for a period of time. For example, looking back to an earlier era, radio took thirty-eight years to hit this benchmark, while the telephone penetrated society in twenty years, followed by the next major innovation, the television, which took thirteen years. Cellular telephones reached the 50 million mark in twelve years, and then the Internet changed the entire dynamic. Within four years of its introduction, the Internet penetrated society. After that, products and websites began to inundate our world, with iPods and blogs taking only three years to hit the magic mark. Then the advent of social media turned the idea of diffusion of innovations upside down. MySpace, the first truly popular social network, required slightly less than two and a half years to penetrate society; it was rapidly overtaken by Facebook, which required only two years to penetrate society (and now, after ten years, has 1.6 billion subscribers, the majority of whom access it daily).[4] YouTube, the popular video-sharing website now owned by Google, took just one year to top the 50 million user mark, and nearly all major websites and apps that followed—including Instagram, Pinterest, WhatsApp, Snapchat, and others—did so in record time. The current hallmark for spectacular penetration is surely held by the smartphone app, Angry Birds, which vaulted from inception to 50 million users in just thirty-five days. Yes, that is not a misprint. Angry Birds became all the rage and penetrated society in just over one month.

Technology has also rapidly invaded our language. Every year the *Oxford English Dictionary* adds new words that it feels deserve to be part of the English language. In recent years the majority of those words revolve around technology or its use, and include terms like "unfriend," "selfies," "hashtag," "tweet," "netbook," "sexting," "cyberbullying," and on and on.[5]

GAME CHANGERS MAKE WAVELETS

As we see it, three major technology breakthroughs have been monumental game changers in our current lifetime: the Internet, social media, and smartphones. By game changers we mean technologies that drive our interference-inducing behaviors—both internally and externally—and which ultimately aggravate our Distracted Minds. Each game changer corresponds to a major invention or trend that drove the wavelets of Toffler's fourth wave and set the stage for us understanding the role technology plays in our twenty-first century Distracted Mind. Some game changers led to a single wavelet while others, such as the Internet, promoted several wavelets, one after the next. First, the web made it possible for anyone to access any information at any time. Second, it enabled email, which made communication virtually instantaneous and free. Third, it gave rise to mobile computing and the ability to access information from any location. Dr. Rosen often refers to the web not as the World Wide Web but as Whatever, Whenever, Wherever, as that is truly how it is used. We no longer need to remember facts, we can simply Google them. In fact, Dr. Betsy Sparrow and her colleagues at Columbia University studied the ability to remember facts and discovered that we were much better at knowing where to find the answers to our questions than we were at remembering the answers themselves. She dubbed this the "Google Effect," and it is indeed a common phenomenon.[6] Just think how many times a day you are in need of some fact and instead of searching your brain to recall whether you know that fact and, if so, unearthing it from memory, you simply press a few keys (or more likely tap a few locations on your smartphone screen) and you have the answer. Even easier, you can ask Siri and she will find the answer for you. Now if you leave a movie theater after seeing a Philip Seymour Hoffman movie and want to know what movie he was in where he played a priest, and who were his costars, the answer is just a few taps away.

The second and third game changers—social media and smartphones—turned our society upside down by creating two new wavelets, one after the other. Let's take a look at each of them separately in terms of their societal penetration and then talk about how they affect our lives.

Social media, at their core, simply use technology to communicate information to many people. Certainly email had been around for quite some time but it is, for the most part, a one-to-one communication. The Internet had already shown through bulletin boards that communication no longer needed to be between two people; it could be among many. However, bulletin boards were typically text based and focused on specific, often technical, topics. Appealing to a wide audience who desired an easy way to talk not one to one, but one to many, social media sites began to emerge. The first viable international phenomenon was MySpace, which was highly graphically oriented and encouraged personalized backgrounds and music selections, providing a palette for self-expression that was quickly embraced by the younger generations. At one point, MySpace, which quickly leapt from obscurity to hundreds of millions of users, was adding 250,000 users a day.[7] But MySpace, as Dr. Rosen said in an earlier book aptly titled *Me, MySpace, and I*, was just too artsy and too "youngish" to appeal to the masses.

Enter Facebook. As most people know from the movie *The Social Network*, Facebook started at Harvard University as a means of connecting university students. To enroll you had to have an email address that ended in ".edu," indicating that you were indeed a university student or at least had a university email account. Quickly saturating that market, Facebook opened to the masses in 2006, and at the latest accounting is now the largest "country" in the world. Strikingly, younger—eighteen- to forty-four-year-old—users check Facebook more than fourteen times a day, spending many hours daily, often broken up into short two minute bursts, reading, commenting, posting, and communicating with their large number of known and unknown "friends."[8]

Where MySpace had a multisensory, graphical, musical, artistic interface, Facebook opted for a more slimmed-down, basic design that encouraged users to accumulate friends and share their lives with them in words, visuals, and "likes." Facebook's history is legendary, and there is no need to document its rapid rise not only in American society but also around the world. One need only look at the use of Facebook and other social media such as Twitter in promoting the Arab Spring to see how this fourth wavelet has become an international one.

While Facebook thrived on an expanded palette that allowed users to post extensive commentaries to their wall and comment exhaustively on others' posts, another social media platform, Twitter, opted for a more streamlined approach, limiting communication to brief blasts of 140 characters or less. Both filled our need to reach out to our community of friends—known and unknown—and communicate at will. And at this writing, many more social media websites have emerged and penetrated society, including photo-sharing sites such as Flickr and Instagram, information-sharing trend sites such as Reddit, video-sharing websites such as YouTube, travel-community-rating websites including TripAdvisor and Yelp, and even online gaming communities such as World of Warcraft and Minecraft. With the rapid rise of these websites, the Internet has truly emerged as a platform for the masses, and as of 2016 the majority of activities are social and informational in nature.

The smartphone revolution that spawned the current Information Age wavelet started with personal data assistants such as the Palm Pilot and first gained universal popularity with the introduction of two devices, the iPhone and the Blackberry. Ostensibly, the Blackberry was a business device, while the iPhone—introduced in 2007—was touted as more of a personal device, or as Steve Jobs said, "a revolutionary product that changes everything." It didn't take long for the idea of carrying a device in your pocket or purse that integrated all of your personal devices into one to take hold. While the Blackberry excelled in providing a portable computer and telephone, the iPhone went several steps further and added in a touch screen, a digital camera, a media player, a GPS navigation system, web browsing, and anything that any of the millions of app developers could imagine. In fact, apps became the trademark of why phones using the iOS or the Android operating system excelled and the Blackberry fell into oblivion. If you could dream of something to make life easier, you could develop an app for a smartphone and hope that the public embraced your vision.

Smartphones are now so ubiquitous that more than seven in ten Americans own one, more than 860 million Europeans own one, and more than half of all cell phone owners in Asia have at least one smartphone if not more.[9] More photographs are taken with smartphones than with digital cameras, and more online shopping is done via smartphones than through standard computers. Statistics show that smartphone users pick up their phone an average of 27 times a day, ranging from 14 to 150 times per day depending on the study, the population, and the number of years that someone has owned the smartphone—those who have owned a smartphone longer check it far more often than those who have recently obtained a phone. Often there is no good reason for them to do so; 42 percent check their phone when they have time to kill (which rises to 55 percent of young adults), while only 23 percent claim to do so when there is something specific they need to do.[10]

Since smartphones allow easy access to information, we have adapted our lives to include them in every part of our day and during everything that we do. Research shows that despite multiple warnings of serious danger (and massive fines), 55 percent of adults access their phone while driving their car, 35 percent use their smartphone in a movie theater in spite of multiple admonitions prior to the movie, 33 percent use one on a dinner date, 32 percent of parents can't resist checking in while they attend their child's school function, 19 percent admit to using theirs in church, 12 percent use theirs in the shower (you can buy a waterproof case if you can't live without your phone while you bathe), and even 9 percent use them during sex.[11] We also appear to be exhibiting a form of bottom-up reflex reaction to our phones, with one in three adults checking their phone immediately after a notification or alert, which rises to four in ten when considering young adults between eighteen and thirty-four. If that's not enough, three in four smartphone users admit to being within five feet of their phone day and night, and 75 percent of teens and young adults sleep with their phone next to their bed either with the ringer on or with the phone set to vibrate. Nearly eight in ten smartphone users reach for their phone within fifteen minutes of awakening, and 62 percent grab their phone immediately upon opening their eyes (regardless of the time of day or night); in the younger set—between the ages of eighteen and twenty-four—those figures skyrocket to 89 percent and 74 percent, respectively. We will talk about the potential devastating effect this has on our mental and physical health in the next chapter.

The web is most certainly Whatever, Wherever, and Whenever, now more than ever. In the mid-to-late 1980s, when desktop computers started to penetrate society, they were most often found in business offices, dens, home offices, or kitchens. With the laptop many moved into the bedroom, but the smartphone sealed the deal. No longer was the bedroom a place to watch nighttime television and sleep. It has now become a haven for multiple technologies. According to Motorola's fourth annual Media Engagement Barometer of 9,500 adults in seventeen countries, video is watched most often in the bedroom on smartphones (46 percent), followed by tablets such as the iPad (41 percent), followed by a television set (only 36 percent).[12] And even when someone is watching an actual television set, he or she is most likely also using at least one other device.[13] Our immersion in high-tech is certainly no surprise to you, the reader, as you are likely to be reading this book on a computer, tablet, or phone, and equally likely to be participating in additional interference-inducing behaviors while you are reading. Let's next explore our penchant for engaging in multiple modern tech devices at a time.

MULTITASKING, TASK SWITCHING, AND CONTINUOUS PARTIAL ATTENTION

Our technology-rich world has proven to be both a blessing and a curse. While on the one hand we have access to information or people anywhere at any time, on the other hand we find our attention constantly drawn by the rich, multisensory, technological environments. It all started with the graphical user interface that took us from the flat, two-dimensional text-based environment that operated on a line-by-line basis similar to a typewriter, to a small picture depicting an operation or program. From there it was a short hop to a completely multisensory world appealing to all of our visual, auditory, and tactile or kinesthetic senses. We now see videos in high definition, often in simulated 3D. We hear high-definition stereo sounds that feel as crisp as sounds in the real world. Our devices vibrate, shake, rattle, and roll, and our attention is captured. It is no accident that we now attach specific ringtones and vibrations to certain people to grab our attention. When Dr. Rosen hears that piano riff from his iPhone he knows it must be either his fiancée or one of his four children, and he grabs the phone before the end of the first few notes. As B. F. Skinner would say, he has been positively reinforced on a fixed-ratio schedule, as it is almost always a positive experience to talk to any of them. On the other hand, several people in his contact list have an "alarm" ringtone, which causes the exact opposite visceral reaction, and he reaches for the button to ignore the call.

Our technology continues to find ways to attract our attention because this is what brings "eyeballs," and the common marketing wisdom is that eyeballs bring money. As you glance at your iPhone you see little red circles with white numbers indicating that something awaits you: four unread email messages, ten Facebook notifications, and so many reminders that your mind is overwhelmed with which icon to tap first. Your iPad does the same, as does your laptop, which particularly taunts you with numerical notifications of unread messages, flashing icons telling you that you need to back up your computer files, and on and on.

Media multitasking—which as described in Part I is accomplished by your brain not performing two tasks simultaneously but instead by rapidly switching from one task to another—occurs in every sphere of our world including home, school, workplace, and our leisure life. *And this is not just limited to the younger generation.* A recent study followed a group of young adults and a group of older adults who wore biometric belts with embedded eyeglass cameras for more than 300 hours of leisure time.[14] While the younger adults switched from task to task twenty-seven times per hour—once every two minutes—the older adults were not all that great at maintaining their attention either, switching tasks seventeen times per hour, or once every three to four minutes. Former Microsoft executive Linda Stone dubbed this constant multitasking "continuous partial attention."[15] As we will discuss later in this book, frequent task switching is something we all do, and the more often we switch, the more detrimental it is to our real-world performance.

Unless you monitor someone's computer as well as his or her smartphone and all his or her other devices, it is difficult to know how much task switching is truly occurring. However, several studies have used different research tools to try to assess real-world task switching. For example, in a recent study Dr. Rosen's lab observed students—ranging from middle school to college age—studying for fifteen minutes in an area where they normally study. Shockingly, students could not focus for more than three to five minutes *even when they were told to study something very important.*[16] This study replicated work by Dr. Gloria Mark and her colleagues at the University of California, Irvine, who observed that IT workers were similarly easily and frequently interrupted.[17]

Other researchers have asked people to keep detailed diaries of their daily media and technology use; one particular study of 3,048 thirteen- to sixty-five-year-old Dutch teens and adults found that people of all ages multitasked at least a quarter of the time—with teens dual tasking 31 percent of their day—although their most common combinations were different.[18] While thirteen- to sixteen-year-olds preferred to combine listening to music with being online, engaging in social media, or viewing online videos, young adults (twenty-five to twenty-nine) preferred combining email, watching television, and visiting websites, and older people (fifty to sixty-five) preferred combining more traditional media activities such as email and radio, television, and visiting websites. Other studies have validated and extended these results; research from Dr. Rosen's lab showed that, when asked how easy or difficult it was to pair a variety of tasks together, members of younger generations reported that they felt that it was rather easy to pair most tasks, while those of older generations felt that only more well-practiced tasks could be easily combined.[19]

One interesting aspect of this penchant for combining tasks is that we seem to have lost the ability to *single task*. Glance around a restaurant, look at people walking on a city street, pay attention to people waiting in line for a movie or the theater, and you will see busily tapping fingers. We act as though we are no longer interested in or able to stay idle and simply do nothing. We appear to care more about the people who are available through our devices than those who are right in front of our faces. And perhaps more critically, we appear to have lost the ability to simply be alone with our thoughts.

Dr. Rosen's lab has been studying this phenomenon for the past decade and has seen a constant increase across generations in how often people check in with their devices. The vast majority of young people check their smartphones every fifteen minutes or less and three out of four young adults sleep with their phones nearby with the ringer on or on vibrate so as not to miss a nighttime alert. While the typical college student owns an average of seven tech devices, older adults are not far behind.[20] Where we used to read, we now skim. Where we used to write, we now use shortened fragments to convey our thoughts. Write a letter? It's much easier to zip off a brief text or an email message. When Twitter first appeared we used to shake our head at the impossibility of putting our thoughts into "only" 140 characters. Now this appears normal and fits our task-switching lifestyle. When was the last time you read a book, a long article, or literally anything more than a page or two without taking a quick peek at your phone or web browser or without the television on in the background? Eye-tracking studies show that when we read a webpage or any text on a screen we don't read it the same way that we read a book.[21] Rather than our eyes passing from word to word along each successive line of text, we tend to read in an "F" pattern, where we read the top and left sides of the page, with a brief foray into the text somewhere in the middle, rather than the complete page line by line. Add in hyperlinks, ads, multimedia videos, scroll bars, and all of the other enticing distractions on a webpage, and it is not surprising that we have difficulty attending to anything for more than a few minutes.

We are most certainly impatient, which you can verify by watching a group of people all checking their phones every three to five minutes regardless of what they are doing at the time or who they are with. A recent study from University of Massachusetts, Amherst, and Akamai Technologies demonstrated our collective impatience by collecting server data from 23 million online video views; the data showed that average viewers begin to abandon a video if it takes more than two seconds to buffer, and 6 percent more viewers click on something else every additional second of buffering.[22] By these data, even a brief ten-second delay in starting a video provokes nearly two-thirds of viewers to leave that screen for another source of information. These quantitative data, collected without the knowledge of the viewers, corroborate survey and experimental data highlighting what was originally dubbed the "four-second rule," referring to the time that an average online shopper is likely to leave a website for another if it does not download.[23] More recent work has even suggested that the four-second rule may actually be closer to a "two-second rule" or even a "400 millisecond rule" (less than half a second), indicating that we are all quite impatient and prone to diverting our attention rapidly from one screen to the next if our needs are not being met instantly.[24] In the next few sections, we take a brief look at research performed in a variety of typical situations where we are prone to interference.

WORKPLACE INTERFERENCE

For those of us who work with technology and are surrounded by other employees working with their technologies, interference has become the norm. We are constantly interrupted by others dropping by our desk to chat or attempting to connect with us through a variety of technological communication modalities, including the most popular workplace tool—email. A study by Judy Wajcman, a sociology professor at the London School of Economics, highlighted this phenomenon by shadowing eighteen employees of an Australian telecommunications company during their entire workday.[25] Wajcman selected this company because it was designed to facilitate interactions between workers with open-plan offices and other external distractors, including many large television screens mounted around the office. The employees in this study spent only half their workday on actual "work episodes," which included any and all work-related activities. Strikingly, most of these work episodes lasted ten minutes or less, with an average of just three minutes per work episode. And even more interesting, nearly two-thirds of the work episode interruptions were self-generated, and most of those involved some form of mediated communication using a technological device. In fact, of the approximately eighty-six daily changes in an employee's work activity, the workers themselves generated sixty-five of them internally, with the vast majority involving "checking in" with no obvious external alert or notification. Even without the "You've Got Mail" notification, these workers checked their email anyway and continued to check other sources of electronic communication and information without being externally directed to do so.

Whether directed externally via an alert or notification or internally by an unseen process, it appears that in the work environment email and other communication modalities bear a major responsibility for interruptions. One field study that followed workers for two weeks discovered that they were interrupted 4.28 times per hour by email and an additional 3.21 times by instant message communications.[26] And these communications appeared to have a strong draw for the employees, since 41 percent of them responded to the email immediately and 71 percent responded to an instant message immediately. On average, the workers spent ten minutes dealing with the alerts and then took an additional ten to fifteen minutes to return to their appointed task, often visiting several other applications in the interim. Another study by the research group ClearContext indicated that more than half of the 250 workers they queried spent over two hours a day reading and responding to email.[27] A study out of Loughborough University in England found that after dealing with an email, which itself took an average of just under two minutes, it took the studied workers an average of 68 seconds—more than half of the time required to read and respond to that email—to return to their work and remember what they were doing.[28] This study also found that people are responding like Pavlov's dogs to incoming email communication, waiting only an average of one minute and forty-four seconds to open that message. Strikingly, 70 percent of those alerts were attended to within six seconds, which is about the time it takes a phone to ring three times. And yet another study found that even without an alert, while one in three people claimed to check their email every fifteen minutes, they actually checked it about every five minutes.[29] We are self-interrupting and not even aware of how often we are diverting our attention from our main task—in this case, our job—to another task that may be completely unrelated to work.

EDUCATIONAL INTERFERENCE

Many studies have examined technology use related to education both in and out of the classroom and its impact on the Distracted Mind. Today's college students own an average of seven high-tech devices, and most students have at least three—smartphone, laptop, and tablet—in the classroom. These devices themselves tend to be used as multitasking tools. Only one in five apps on college students' smartphones were categorized as "productivity" apps.[30] In the classroom, these devices provide a ready source of interruption that has been validated in many studies. For example, one study found that nine in ten students used their laptop computers for nonacademic reasons during class time, while another study found that 91 percent of students reported texting during class.[31]

Other studies have addressed how students use technology while they are studying outside the classroom. Terry Judd, a professor at the University of Melbourne, monitored more than 3,300 computer session logs from 1,229 students studying in the computer lab and found that the average time on task was only 2.3 minutes; multitasking was the name of the game, with less than 10 percent of the sessions being devoid of task switching to something other than studying, which turned out to be primarily checking email, texting, and social media.[32] In a laboratory study, a researcher from Virginia Commonwealth University observed college students during a three-hour study session using video cameras and eye trackers and found that on average, students spent more than an hour listening to music and showed thirty-five interruptions of six seconds or longer, totaling twenty-six disrupted minutes in just three hours.[33] The biggest cause of interruptions was the smartphone, which students checked close to nine times in the three-hour study session. Other major interrupting culprits included checking the Internet for information not related to the material being studied and checking email.

Another report on the studying activities of students found that the reason behind the constant task switching is a desire to feed emotional needs—often by switching from school work to entertainment or social communication—rather than cognitive or intellectual needs.[34] According to the study's authors, "This is worrisome because students begin to feel like they need to have the TV on or they need to continually check their text messages or computer while they do their homework. It's not helping them, but they get an emotional reward that keeps them doing it."[35]

More work has been done to document the impact of technology on student behavior than any other population, as technology is more readily available to these individuals and they are the first to have grown up immersed in a technology-rich environment with ever-increasing opportunities for interference. Recall the study introduced earlier in this chapter where middle school, high school, and university students were observed while they were instructed to study *something important* for a short period of time (only fifteen minutes).[36] Regardless of age, students were able to stay focused and attend to that important work only for a short period of time—three to five minutes—before most students self-interrupted their studying to switch to another task. During the fifteen-minute study period, students were able to actually study for only nine minutes. The major culprits that spurred the constant interruptions had two sources: social media and texting. Both of these were apparently offering such important information that the studying student's attention was transferred from the task at hand—an identified important area for focus—to another source of information through the two most popular communication modalities among the younger generations.

EVERYDAY MEDIA MULTITASKING

A recent article has described the concept of "everyday media multitasking" as including task switching during normal daily activities.[37] One of the interesting aspects of everyday multitasking has to do with perceptions of how easy or difficult it is to multitask with pairs of activities. As part of the study mentioned earlier in this chapter, Dr. Rosen's lab found general agreement across generations on which tasks were easier or more difficult to perform together.[38] All generations indicated, for example, that it was difficult to combine playing video games with other tech activities. However, members of younger generations—this study compared baby boomers (born between 1946 and 1964), Generation X (1965–1979), and the Net Generation (1980–1989)—perceived vastly more tasks to be easily combined than did members of older generations. For example, while only 39 percent of baby boomers felt that it was easy to both email and text, 63 percent of Gen Xers and 81 percent of Net Geners reported that it was easy for them to do both at the same time. This pattern was evident for nearly all combinations of paired technological tasks.

The study that compared ease or difficulty of everyday multitasking across generations was originally performed in 2008.[39] In 2014, Dr. Rosen's lab replicated the study with a new sample including members of four generations—baby boomers, Generation X, the Net Generation, and the iGeneration (born between 1990 and 1999). We were interested in seeing whether people of different generations were attempting, six years later, to multitask at the same level. Participants were asked whether they regularly attempted to perform each of sixty-six task pairs at the same time, including simultaneously combining two tech activities (e.g., playing a video game and texting) and also combining a tech task with a non-tech task (e.g., being online and eating). In total there were nine tech tasks and three non-tech tasks (eating, talking to someone face to face, and pleasure reading). In 2008, baby boomers attempted 59 percent of the task pairs. In 2014, that number had risen to 67 percent. Generation X increased their attempts from 67 percent to 70 percent, while those in the Net Generation increased from 75 percent to 81 percent. Strikingly, iGeneration teens and young adults claimed that they attempted to combine 87 percent of the pairs. It appears that in a few short years we are seeing even more attempts at multitasking, which we now know is not the most effective way of performing tasks successfully.

The most often reported task that gets combined with other tasks turns out to be the one media form that is used by more people of every generation: the television. Dubbing this the "second screen" problem, a study by the Council for Research Excellence (CRE) discovered that among 3,000 American teens and adults who watched television on a variety of devices (the new norm), 55 percent of the time they were watching on a television set they were using a second screen.[40] Participants were also using second screens 61 percent of the time they were watching TV on a computer, 59 percent of the time they were watching on a tablet, and 53 percent of the time they were watching on a smartphone. Some in the entertainment field have even encouraged viewers that TV watching should be combined with social media, so that you watch a show and comment on it as it unfolds. Anderson Cooper, for example, routinely informs his CNN audience that he will be tweeting, blogging, and posting on social media and asks them to join in with him. However, the CRE study found that nearly two-thirds of those second-screen activities were unrelated to the show content. A laboratory study in Belgium validated and extended these findings to demonstrate how young adults watch news programs.[41] Requiring viewers to respond to second-screen tasks, some of which were relevant to the news stories and some of which were not, the researchers discovered that regardless of the relevance of the second-screen activity, factual recall of the news show was reduced when splitting attention.

Another laboratory-based study found that when given the option of watching television and doing other activities on a computer, young adults switched between the two screens four times a minute for a total of 120 switches while viewing a thirty-minute video.[42] Using glasses that tracked eye movements and fixations, the researchers discovered that gazes at the television were substantially shorter (1.8 seconds) than those at the computer (5.3 seconds) and that the young adults, when asked immediately after the study how often they felt they switched between the computer and the television, recalled only 12 percent of those switches. Interestingly, a similar study by Comscore suggested that second screens enhance attention to and engagement with entertainment TV programs, which possibly suggests that second-screen activity while trying to watch "factual" television may be more disrupted than using additional media resources when watching entertainment television.[43]

A recent headline proclaimed "4 in 5 Americans Multitask While Watching TV" and reported a Deloitte study showing that 81 percent of Americans across generations almost always or always engage in at least one more activity when watching television; for young adults this figure approaches nearly 90 percent.[44] Deloitte separated their report on this second-screen phenomenon into generations and found that even 63 percent of "matures" (sixty-six and older) and 77 percent of boomers (forty-seven to sixty-five) multitasked while watching TV, mostly checking email and surfing the Web. Eighty-seven percent of Gen Xers, 88 percent of "Leading Millennials" (twenty-four to twenty-nine), and 86 percent of "Trailing Millennials" (fourteen to twenty-three) engaged in nearly every possible activity while simultaneously watching television. Another recent study set in the UK found that when 200 people were asked to track their media activities during a one-hour evening period, they reported switching between mobile phones, tablets, and laptops twenty-one times, and 95 percent of them did this with the television on for the entire hour.[45]

With television itself a multitasking environment complete with scrolling information and more possible viewing options on other portions of the screen, it appears that we still feel the need to augment that high-definition, multisensory experience with ever-increasing amounts of information and communication. As mentioned above, some of this concurrent activity is not for purposes of distraction but rather to augment the viewing experience by accessing social media to comment on the content of what we are watching on television. Results of one study that compared the experience of simply watching television with that of watching and tweeting at the same time suggested that adding in a social presence actually enhanced the experience and decreased the boredom of simply watching a television program by itself.[46] It is fascinating that early television viewers were warned that the set may be overstimulating, and now we appear to be consciously adding in other media for the express purpose of escaping apparent "understimulation."

Another everyday multitasking issue involves bringing work from the office to other locations. Two national studies, one performed in the United States and the other in Canada, found that the vast majority of workplace professionals in both countries bring work home to complete after normal business hours.[47] It doesn't take much imagination to guess what work they are doing at home: emails and Internet searches top the bill. Research also reveals that smartphone users can't seem to bear being separated from their devices. Another study that surveyed 3,600 young adults eighteen to thirty years old from eighteen countries found that three out of four check their phones in bed, more than a third check it in the bathroom, nearly half check during meals, and four in ten admit to feeling anxious when they are apart from their phone.[48] A Bank of America study found that 47 percent of US adults admit they wouldn't last a day without their smartphone.[49] And these are adults! Internet gaming addiction was included in the appendix of the new fifth edition of the *Diagnostic and Statistical Manual of Mental Disorders*, the bible for diagnosing specific psychological syndromes. Perhaps by the time they get around to producing the sixth edition, Internet gaming addiction will be replaced by smartphone addiction.

Finally, remember taking a vacation before smartphones—and worrying about having to deal with hundreds of emails when you returned home? Perhaps you even felt compelled to visit an Internet cafe, to log in and check your messages midway through your trip. Now, eight in ten smartphone users take them on vacations and check in frequently during the day and night.[50]

EXPECTATIONS

One area in which technology has had a major impact—and which has exacerbated the impact of interference—is in our expectations of other people's behavior and of our own behavior. An early example is the standard telephone. When telephones had successfully penetrated society and became a common way to communicate with family and friends, our expectations were clear: if the phone rang and rang and rang and nobody answered, then we knew that we had to try to call again later. When answering machines became available and were positioned next to the phone, the expectation was that a message left during the day would likely be heard when the receiver arrived home (assuming he or she checked the machine and noticed the number indicating awaiting messages), and a return call would likely come then or perhaps the next day if the hour was late. Then the answering machine moved into cyberspace in the form of voice mail: now you had to call your phone, press the * key (or some special key combination), enter your code, and hear your messages. The voice mail system did not necessarily alert you to the message, so you had to call in periodically to check (unless a light flashed on your phone signifying a voice mail). This system was all quite simple and quite civilized, as our expectations had boundaries.

When cell phones and later smartphones came blasting into our world, our expectations changed. "Sorry, I wasn't home to get your call" or "I didn't have a chance to check my messages" were no longer excuses when the phone icon showed you that you had missed a call and had a voice mail waiting. We now feel compelled to answer calls or respond to voice mail immediately, regardless of our circumstances. A phone call while dining might elicit a sheepish response from our dining companion of "Sorry, but I have to take this call." It was not long ago that if you heard someone muttering behind a stall door in a public restroom you assumed he was talking to himself or, in the extreme, might be facing psychological issues. Now, it is commonplace to hear people talking on their phones in restrooms, in churches, and just about anywhere you can imagine.

The same radical change in our expectations has extended to all communication modalities. If our text message is not responded to immediately, we attribute motivations to the recipient—"She must be mad at me"—or if we comment on someone's Facebook post and he doesn't immediately respond or, at the least, "like" our comment, we get miffed and feel rebuffed. As more of our personal communications move from the real world to the virtual world, more opportunities arise for others to not meet our expectations. And this is not just about personal communication. In the workplace, the expectation is that a communication *must* be addressed immediately. And those expectations mount with electronic conversations between more than two people. If, for example, your manager sends a group email, once the first person responds the clock is ticking; if you do not jump on the email chain and chime in, you are not meeting the new norms of instant communication. It doesn't matter if you are working arduously on a project for that very same boss; the expectation is that you immediately cease what you are doing—clearly interfering with your train of thought—and respond, lest you be judged negatively. This is a main reason why, as mentioned earlier when discussing workplace interference, nearly half of workplace employees respond to an email immediately and then spend ten minutes dealing with the content of that email, only to take an additional ten to fifteen minutes to return to their work.[51]

Finally, the introduction of electronic communication modalities into the workplace has changed more than just our workday expectations. Employees, for the most part, are now expected to respond to afterhours communiqués as rapidly as they do to workplace messages. Essentially, the workplace has become, owing to our new response time expectations, a 24/7 experience. And, as we have said before, even vacations do not allow us to escape from being "always on, always available."

In this chapter, we have examined how technology has, in just a few short years, become a main staple of our personal life, our education, our workplace, and our personal interactions with those around us. It is the driving force behind our incessant quest for information. Technology is precisely what has driven the major changes in our world and produced a rapid series of technology wavelets that are constantly tugging at our abilities and our perceptions of the world. We have shown through solid research how and why our relationship with our devices interferes constantly in our daily lives. Note, however, that technology did not create the Distracted Mind; but it did aggravate the conflict between our high-level goals and our cognitive control limitations, discussed in Part I, which underlies the Distracted Mind. Next, it is important to determine whether this is simply a natural, evolutionary change in how we navigate our current world—with little or no impact on our productivity, performance, and relationships—or whether interference, each clarion call from our brain to switch to another task and each vibrating alert and notification drawing away our attention, is having deleterious effects on our performance, our relationships, and our physical and mental health.

7 THE IMPACT OF CONSTANTLY SHIFTING OUR ATTENTION

IN CHAPTER 6, we documented how modern technology, while providing major benefits in terms of information flow and rapid communications, has aggravated our already Distracted Mind and put us in a situation where we have seemingly lost the ability to control our use of that technology. We can no longer focus in the classroom or the workplace; nor can we resist the pull of responding to alerts and notifications while we are supposed to be spending time with our family and friends. We have adopted a style of "everyday media multitasking"—which is just a generous way of saying that we have lost our awareness of what is necessary and what is simply reflexive responding as though prodded by a sharp stick. In this chapter, we will discuss the consequences of this constant goal interference for our real-world, everyday performance in the classroom, in the workplace, on the road, and more.

HIGHER EDUCATION

Current university students face a major challenge. Unless they are returning older students, they are members of either the iGeneration or the Net Generation whose childhood was replete with every form of media and technology available, and they spent their time attempting to multitask at every turn. When they enter college with, as we documented in chapter 6, a penchant for engaging in multiple sources of information—what we might think of as constantly juggling chainsaws—if they cannot focus on their studies, the consequences are great. Education takes place in two locales: in the classroom and outside the classroom. In this section, we will first present the evidence for the negative influence of high-tech on how students study and then follow with the ramifications of their inability to pay attention during class sessions.

In one study mentioned earlier, Dr. Rosen's research team observed hundreds of middle school, high school, and university students studying something important for fifteen minutes in the environment where they normally study.[1] Minute-by-minute observations showed that the typical student couldn't stay focused on work for more than three to five minutes. As a measure of school success, students were asked to provide their grade point average (GPA), which was on a consistent four-point scale regardless of school level. Strikingly, the predictors of a lower GPA from extensive data collected about the students were: percentage of time on-task, studying strategies, total media time during a typical day, and preference for task-switching rather than working on a task until it was completed. In addition, by examining the websites that students visited during that fifteen-minute sample, we uncovered a fifth predictor of a lower GPA. Only one website visited predicted a lower GPA: Facebook. And it did not matter whether the students visited it once or fifteen times. Once was enough to predict lower school performance.

When university students attend to interruptions from their vast array of media devices while studying, one ramification appears to be an increase in the amount of time it takes them to study compared to simply studying without interference. In a laboratory experiment by Laura Bowman and her colleagues at Connecticut State University, students were randomly assigned to three groups to read a book chapter and take a test.[2] One group simply read the chapter and took the test. The second group first completed an instant message conversation with the experimenter and then read the chapter and took the test. The third group started to read the chapter, were interrupted with the same instant message conversation, which was delivered in pieces at various times during the reading, and then took the test. This latter group was designed to simulate the typical studying behavior of college students. The interesting results showed that all groups performed equally well on the test, but the third group took substantially longer even when the time spent instant messaging was removed. If you visualize a college student studying in the dorm and constantly checking his or her smartphone or social media, then you soon get the idea that the constant attempts to multitask tax all aspects of cognitive control and lead to many all-nighters or at best constant late-night study sessions.

An interesting study out of Washington State University attempted to pinpoint the consequences of a variety of technology uses while studying and discovered that the top three activities of studying students were listening to music, texting, and social media. The students were asked to report on their "mobile phone interference in life" (MPIL), which included items related to school work ("I would be more productive if I didn't use my mobile phone so much") and other items related to cell phone addiction or overuse ("I have tried to cut down the amount of time spent with my mobile phone, but failed").[3] Only texting and social media while studying were correlated with higher MPIL. As the authors predicted, music did not have the same association. In other words, students who texted or used social media while studying reported that their mobile phone interfered in their lives more than students who listened to music or did not multitask while studying. The authors propose that "the difference between the effects of passive listening to music and active engagement in texting or social media highlights a major shift in the intrusion of media in everyday life. Traditional media, such as radio, television or music, which can be ignored as background noise, are fundamentally different from human interactions via text message or social media."

Clearly, students' cognitive control abilities are being challenged while they are studying. They are not able to focus on a single task for more than a few minutes, notably on the somewhat more boring task of reading (attentional sustainability) especially when more interesting activities abound (attentional selectivity), and they require more time to redirect their efforts after each interruption, resulting in additional time needed to read than if they had simply read straight through with no interruptions (task switching). In addition, their working memory may also be compromised, as distractions degrade the fidelity of the information that they are trying to maintain during the learning process.

As you might expect, technology-related multitasking in class has a negative impact on students' school performance similar to that of multitasking while studying. This has been studied extensively, with researchers linking nearly every type of in-class technology use—including email, texting, laptop, social media, and more—to decreased classroom performance, regardless of how that performance is measured (grades, work productivity, etc.), and across all grade levels ranging from elementary school to college.[4] Studies have shown that excessive multitasking while learning increases the time needed to effectively master the material and also increases the level of stress felt by the learner.[5] Further studies have linked "attempted" classroom multitasking to poor classroom performance.[6] One interesting empirical study by Eileen Wood and her colleagues compared student learning as a function of their classroom multitasking by requiring students to use a specific technology—social media, texting, emailing, or instant messaging—while attending three course lectures. All four multitasking groups, plus a group that was allowed to use any technology during the lectures, performed worse on a test of the delivered content than those who did not use any technology during the lectures.[7]

Another study validating the negative impact of classroom multitasking interrupted students during a short video lecture and required them to either text the experimenter or post material on social media, under two conditions: one new text or post every minute, or one new text or post every thirty seconds (chosen by the authors to simulate normal in-class texting experiences of many iGeneration students).[8] A control group simply watched the video, which was followed by a test. The results of this study found that more texting or social media posting resulted in poorer lecture notes and lower test scores than the control group. In addition, a negative linear trend emerged in both lecture notes and test scores, with the highest scores and best notes demonstrated by those students who did not receive any interruptions, followed by lesser scores and notes of students who were interrupted every minute, and, finally, the worst scores and notes of students who were interrupted every thirty seconds.

According to many research studies, most students send and receive text messages during class, and those who do get lower grades.[9] In a study by Dr.

Rosen's lab, students were sent varying amounts of text messages at crucial points of a videotape lecture and asked to respond.[10] Those who received eight text messages during the thirty-minute lecture performed an entire grade lower in a test of the lecture material than the average of those who received no texts or only four texts. Interestingly, according to another study, students are aware of the potential downsides of such interruptions.[11] When asked to estimate how they would do on a test if they were to text during a lecture, they guessed that they would perform 30 percent worse than if they didn't text and, indeed, they scored 30 percent worse. But even with that knowledge, they do not change their behavior. With the continuing trend of incorporating technology in the classroom, these data suggest that caution may be in order.

In addition to examining the negative impact on their course performance, several research studies have shown even more far-reaching effects of technology use by college students. One study found that those students who used cell phones and texted more often during class showed more anxiety, had lower GPAs, and were less satisfied with life than students who used phones and texted less frequently.[12] As we will discuss in chapter 9, anxiety may be driving the need for young adults to constantly check in with their technology regardless of where they are at the time. Another study found that when freshman students have more Facebook friends, they tend to be less emotionally and academically adjusted, whereas junior and seniors with more Facebook friends are generally more socially adjusted and attached to their institution. The authors propose that the shift in relationship between Facebook and well-being may be due to upper-class students using Facebook to connect socially with their peers and participate in college life.[13] Finally, a study of more than 770 college students discovered that students who used more interfering technology in the classroom also tended to engage in more high-risk behaviors, including using alcohol, cigarettes, marijuana, and other drugs, drunk driving, fighting, and having multiple sex partners. Overall, it appears that college students who use inessential technology either during class sessions or while studying face difficulties on both an academic and personal level.[14]

SAFETY

When college students switch from studying at home or attending a lecture in class to some other nonessential activity, the ramifications are certainly important, but not life threatening. Other examples of interference can have dangerous consequences. In this section we will explore some critical situations that can arise from our Distracted Mind, and highlight the role of technology. One compelling example of our limitations in attention involves what is called "attentional blindness." This occurs when your top-down control focuses your selective attention to such a degree that you are not aware of bottom-up stimuli, even stimuli that may normally seem so novel or salient you would expect to notice them. You may have heard about the "invisible gorilla experiment" where a video is shown of two teams who are passing a basketball between their teammates, one wearing white and one wearing black. In the classic version, study participants are asked to watch the video and count the number of passes between members of the white-clad team. Somewhere in the middle of the video a woman dressed in a black gorilla suit walks between the players, beats her chest, and moves out of the video scene. After the conclusion of the video, participants are asked how many passes they counted. Most get that part right. Then they are asked, "Did you see anything out of the ordinary happen in the video?" and in most studies roughly half do not report seeing the gorilla at all.[15]

Ira Hyman and his colleagues incorporated an interesting twist on the gorilla study at Western Washington University.[16] Instead of the gorilla, which might be lost in the crowd of basketball passers since it is wearing black as are half the basketball players, Hyman had a clown, fully clothed in a bright purple and yellow outfit, with large shoes and a bright red bulbous nose, pedal a unicycle around a large open square that is traversed often by most campus students during a typical day. The researchers interviewed more than 150 students who walked through the square and noted if they were walking alone or with someone else and if they were using a cell phone or listening to music with ear buds. When asked if they saw anything unusual, only 8 percent of cell phone users reported that they saw the clown. This was compared with one in three students walking alone without technology or listening to music wearing ear buds and more than half of he students who were walking in pairs without using technology. When asked directly if they saw a clown, still only one in four of the cell-phone-using students reported seeing it compared with half of single walkers, 61 percent of music listeners, and 71 percent of walking pairs. Whatever was happening between the user and his or her phone appears to have inhibited their ability to identify such a strong bottom-up event in the immediate neighborhood. Imagine a squirrel so involved in eating his acorns that he totally ignores a predator, and you have the unlikely parallel to this occurrence in the animal world.

The public became more broadly aware of the impact of inattentional blindness on safety due to technology use in a humorous way in early 2011, when Cathy Cruz Marrero was walking in a mall while texting on her phone and fell headfirst into a fountain.[17] The video has been viewed millions of times and made every news show. Although she was not hurt, the event demonstrated how multitasking with technology while walking could be hazardous to your health. According to one report in *Scientific American*, data from a sample of 100 US hospitals found that while in 2004 an estimated nationwide 559 people had hurt themselves by walking into a stationary object while texting, by 2010 that number topped 1,500, and estimates by the study authors predicted the number of injuries would double between 2010 and 2015.[18] A recent study by Corey Basch and her colleagues at several eastern universities tracked more than 3,700 pedestrians crossing Manhattan's most dangerous intersections and discovered that nearly 30 percent focused their attention on their mobile device while crossing during the "walk" signal, and one in four were even looking at their phones while crossing during the "don't walk" signal.[19] While more than half were wearing headphones and presumably listening to music or talking on the phone, substantial numbers were simply using their device for some activity and staring at it instead of watching for traffic on a street known for many pedestrian-motorist accidents. And lest you think this is a New York phenomenon, a group of researchers at the University of Washington and Seattle Children's Hospital found similar results in their study of more than 1,100 pedestrians.[20] They discovered that 30 percent were doing something other than just walking while crossing an intersection, including listening to music and texting. The texting pedestrians took an additional few seconds to cross the street and were nearly four times more likely to show at least one unsafe crossing behavior than those who did not have their head down to look at their phone.

Recent research has shown that simply using a phone while walking modifies the walker's step width, toe clearance, step length, and walking cadence and makes him or her more prone to injuries even without an automobile involved.[21] Direct observational research of pedestrians at a variety of intersections indicates that around three in ten walkers are occupying themselves with their cell phones as they cross the street and that they are nearly four times more likely to behave unsafely while crossing (veering out of the crosswalk, crossing before or after the signal, etc.) than those who are not using any technology as they cross the street.[22] In one experimental study, college students were asked to cross the street in a virtual environment either talking on the phone, listening to music, or texting.[23] Those who were texting or listening to music were more likely to be hit by a simulated car, which the authors attributed to the conflict between the cognitive demands of crossing the street and paying attention to vehicles and the demands of paying attention to the text message conversation or their music—a classic interruption cost, but a deadly one in this case.

While Ms. Marrero's injuries sustained while walking and using a smartphone were minor, if you replace "walking" with "driving" the situation becomes much more serious. The Centers for Disease Control (CDC) reported on their website that in 2011, 3,331 people were killed nationwide and 387,000 were injured in motor vehicle crashes involving an inattentive driver.[24] The US National Safety Council estimated that 23 percent of *all* car crashes, which includes crashes caused by inattention as well as another approximately million crashes for other reasons, involved cell phone use.[25] This is such a serious issue that Matt Richtel of the *New York Times* won a 2010 Pulitzer Prize for Journalism for his series of articles on this topic called "Driven to Distraction."[26]

David Strayer, a professor at the University of Utah and an expert on the impact of technology on driving, compared cell phone drivers and drunk drivers and discovered that a person using a cell phone while driving and a person with a blood alcohol level above the legal limit *have an equal chance of being in a traffic accident.*[27] According to the CDC, 69 percent of adult drivers in the United States reported that they had spoken on their cell phones while driving within the past thirty days. When the CDC examined comparable statistics from other countries, the percentages of cell phone use while driving ranged from 21 percent in the UK to 59 percent in Portugal. In the same report, the CDC noted that 31 percent of US adults reported texting and driving, again higher than drivers from any other country. When looking at data from high school students, the CDC found that nearly half text or email while driving. This is in spite of the fact that most states have laws—and quite large fines—for using a handheld cell phone or texting while driving.

Driving an automobile is one area that requires extensive use of all aspects of cognitive control—attention, working memory, and goal management. According to David Strayer, "digital distraction is a significant source of motor vehicle accidents."[28] Cell-phone-related accidents can happen for a variety of reasons, including taking one's hands off the steering wheel, taking one's eyes off the road, or withdrawing one's attention. Given that research shows that the incidence of accidents with handsfree phones and handheld phones are equivalent, it is likely that the primary cause is neither physical nor visual but rather an issue of attention—one of our major cognitive control abilities with distinct limitations.[29] Despite this, handsfree cell phone use is still legal in every state.

One further comment needs to be made about voice texting. With a Bluetooth device, various apps can be used to summon a virtual assistant who will read incoming messages and send outgoing ones. Research performed by the American Automobile Association found that "common voice tasks are generally more demanding than natural conversations, listening to the radio, or listening to a book on tape," supporting the conclusion that this approach still uses significant amounts of our limited attentional resources.[30] In fact, the literature on the peril of using handsfree phones applies equally to handsfree messaging aids. By their very nature they require a variety of cognitive control resources, and thus distract you from the road even when your eyes are pointing straight ahead.

Interestingly, when Strayer and his colleagues also investigated the impact of having a conversation with a passenger they found no decrement in carrying out driving tasks, suggesting that talking to a passenger still allows the driver to attend to driving.[31] In fact, Strayer and his colleagues analyzed the driver–passenger exchanges and discovered that, for the most part, the passenger either reminded the driver to perform the task of exiting the highway at a rest stop or simply stopped talking as the decision point approached, allowing the driver to attend to the driving task alone. As the authors conclude, "In effect, the passenger acted as another set of eyes that helped the driver control the vehicle, and this sort of activity is not afforded by cell-phone conversations."

The statistics on technology and driving are alarming, particularly among younger drivers who, as we have discussed earlier in the book, do not yet possess a fully developed cognitive control system and even when not behind the wheel suffer from inattention and poor goal management. Studies have shown that drivers are actually just driving and not doing any other activity only 46 percent of the time. According to studies from the Pew Internet & American Life Project, 26 percent of teens self-report texting while driving and 48 percent report having ridden in cars with texting teen drivers.[32] Elsewhere it has been reported that large numbers of teens are driving and using technology.[33] Texting while driving is responsible for at least 18 percent of deaths due to car accidents and a whopping 24,000 accidents as of 2009, according the National Highway Traffic Safety Administration.[34]

Over the past few years we have seen horrifying repercussions of technology use. A Florida trucker admitted that he was texting just prior to crashing into a school bus full of children. An Arizona trucker was on Facebook just before he crashed into three police cars and two fire trucks responding to a roadside accident. A quick search online shows similar incidents in nearly every state—including train conductors, truckers, and automobile drivers. In each case the driver's attention was diverted, most often to a smartphone, for just enough time to result in a major accident. In addition, reports have been filed of air traffic controllers texting while working and even pilots texting while flying helicopters. Clearly an intersection of modern technology and the Distracted Mind is making us unsafe.

WORKPLACE

In the previous chapter and throughout the book, we have documented the strong lure of interference in the workplace. Countless studies have shown that office workers face constant interruptions from technology to which they respond rapidly, which then results in additional work time to return to the interrupted task. It is not just the extra time it takes to return to work that may cause problems for the office worker. A study of more than 200 employees at a variety of companies investigated the factors that predicted employee stress levels.[35] While having too much work to do was the best predictor, it was only slightly stronger in predicting exhaustion, anxiety, and physical complaints than outside interruptions, many of which were electronic in nature. Other research in the workplace, including the seminal work of Gloria Mark and her colleagues at the University of California, Irvine, has observed information workers and university students performing their normal daily tasks and demonstrated that while interruptions occurred about every three minutes, and for the most part lasted only a short time, it took nearly a half hour for the individual to return to an interrupted task as other activities were pursued.[36] In spite of this, workers actually accomplished their work faster after the interruptions, but that speed came at a mental cost. In summarizing one study, Mark reported that "working faster with interruptions has its cost: people in the interrupted conditions experienced a higher workload, more stress, higher frustration, more time pressure, and effort. So interrupted work may be done faster, but at a price." In a *New York Times* interview, Clive Thompson summed up research results on workplace interruptions by asserting that "we humans are Pavlovian; even though we know we're just pumping ourselves full of stress, we can't help frantically checking our e-mail the instant the bell goes ding."[37]

Eliminating constant interruptions is particularly difficult in an "open office" setting where there are no private offices, but instead employees work in cubicles where the chance of external interruptions increases astronomically. Approximately 70 percent of US offices—including Google, Yahoo, Goldman Sachs, and Facebook—have either no partitions or low ones that do not make for quiet workspaces.[38] Research has shown that open offices promote excessive distractions.[39] In one study, researchers simulated four open-office conditions with various levels acoustic factors and found that background noise generated subjective impressions of a negative impact on performance and quantifiable negative effects on short-term memory and working memory tasks.[40] Another study examined 1,241 employees from five organizations with different office types and concluded that when workers require concentration to perform their job they report more distraction and more cognitive stress in open-office settings.[41] Finally, a content analysis of twenty-seven open-office studies identified auditory distractions, job dissatisfaction, illness, and stress as major ramifications of this type of workplace.[42]

To sum up, the research done in the workplace suggests that we are trying to cram too many tasks into too short a time because we respond as though an alert—usually about an incoming communication—is a command to drop whatever we are doing and reorient our attention to a new patch of information. In addition, while we are usually able to accomplish this (and even faster, according to Gloria Mark's work), we pay a huge price in terms of anxiety and stress. It is important to point out that none of the research studies that examined workplace interruptions dealt with the addition of the smartphone to our vast array of desktop technology. The bottom line is that being constantly interrupted and having to spend extra time to remember what we were doing has a negative impact on workplace productivity and quality of life. One 2005 study—before the major increase in smartphone usage—estimated that when office workers are interrupted as often as eleven times an hour it costs the United States $558 billion per year.[43]

RELATIONSHIPS

One of the most difficult areas to study is the impact of our Distracted Mind on relationships with our family and friends. In her 2011 book titled *Alone Together: Why We Expect More from Technology and Less from Each Other*, Dr. Sherry Turkle, an MIT professor and one of the pioneers of the study of our relationship with technology, argues that our Distracted Mind is removing us from "real connections" and offering only "sips of connection."[44] Turkle sums up her view of the negative impact that technology has on our attention and our important relationships, saying, "As we distribute ourselves, we may abandon ourselves." She further talks about the negative impact that she sees in parenting: "Young people must contend with distracted parents who with their BlackBerries and cell phones may be physically present but 'mentally elsewhere.'"

We see it all around us. Friends sit "alone together" absorbed by the virtual world inside their smartphones and paying little attention to those in their real world. Parents take their children to the park only to spend the entire time reacting to alerts and notifications on their phone instead of engaging with their children in all-important free-play activities. Spouses, who used to watch television together and discuss what they saw and learned, now use a second screen while they attempt to divide their attention between their tablet, phone, or laptop, the content of the television, and their loved one. Something has to suffer, and evidence is starting to indicate that it is our relationships with each other that may be one peril of our Distracted Minds. A 2014 Pew Internet & American Life Project report found that one in four cell phone owners in a marriage or partnership felt that "their spouse or partner was distracted by their cell phone when they were together."[45]

We have probably all experienced a dinner out or even one at home where the table is littered with smartphones. This has become such a concern of young adults that some play a game called "cellphone stack" where everyone at the table places their phone in the center of the table, one on top of the other, and whoever looks at their device before the check arrives must pay the entire bill.

An interesting study by Andrew Przybylski and Netta Weinstein of the University of Essex examined the impact of simply having a phone present during an interpersonal social setting.[46] In two face-to-face studies, the researchers had two people who had never met spend ten minutes either having a casual conversation or discussing meaningful personal matters. In one condition, a mobile phone—not belonging to either of the participants—was placed either on a nearby table within full view but not in the direct line of sight of either one, or was absent and replaced by a similar-size notebook. Following the short conversation, both partners rated their feelings of closeness, trust, empathy, and understanding with the other partner. Strikingly, as the authors concluded, "the mere presence of mobile phones inhibited the development of interpersonal closeness and trust, and reduced the extent to which individuals felt empathy and understanding from their partners." This study was validated by a research team studying the "iPhone Effect" where, using a similar design as the previous study, they compared partners who did not place their own mobile device on the table or hold one in their hands with those who did and found that conversations between these strangers in the presence of a device were rated as less satisfying and were reported as generating less empathic concern.[47] Yet another similar study by researchers at the University of Southern Maine found that "simply the presence of a cell phone and what it might represent (i.e., social connections, broader social network, etc.) can be similarly distracting and have negative consequences in a social interaction."[48] If our Distracted Mind can negatively affect social connections and feelings of closeness just by being in the presence of modern technology during a short conversation with a stranger, what does that imply about how it can impair our real relationships?

MENTAL, EMOTIONAL, AND PHYSICAL HEALTH

At this point in the book, it should be clear that we all possess a Distracted Mind. Thus far we have shown how this interacts with modern technology to affect our cognition, behavior, safety, relationships with family and friends, and performance in various situations, including the classroom and workplace. It doesn't end there; our relationship with technology has spawned a variety of "conditions" that include phantom pocket vibration syndrome, FOMO (fear of missing out), and nomophobia (fear of being out of mobile phone contact), all of which are centered on a need to be connected constantly. Phantom vibrations are an interesting phenomenon. A mere ten years ago, if you felt a tingling near your pants pocket you would reach down and scratch the area to relieve the presumed itch. Now the very same neuronal activity promotes a need for us to check our smartphone—sometimes even if we are not carrying one in our pocket at the time—as it is assumed that our phone just vibrated, signaling an incoming alert or notification. Two studies have discovered that nearly everyone experiences these false vibrations often.[49]

A study performed by Dr. Rosen's lab of 1,143 teens, young adults, and adults assessed symptoms of psychiatric disorders, daily media and technology use, preference for multitasking, anxiety about missing out on technology use, and technology-related attitudes.[50] Overall, symptoms of psychiatric disorders were predicted by some combination of daily technology use and preference for multitasking even after factoring out the impact of anxiety about missing out on technology and technology-related attitudes.

Another recent study from Dr. Rosen's lab spearheaded by Dr. Nancy Cheever investigated the role that technology use—or, rather, lack of use—has on anxiety.[51] One hundred sixty three college students were brought into a lecture hall, with half being told to turn off their phone and store it and all other materials under their seat while remaining quiet and simply doing nothing. The other half of the students were given the same general instructions about storing materials out of sight and doing nothing, but they had their smartphones taken away and replaced with a claim check for later retrieval. Ten minutes later and then twice more during the hour-plus session, each student completed a paper-and-pencil measure of anxiety. The prediction was that the students holding a claim check for their phone would become anxious, and indeed they did—but no more so than the students whose phone was turned off and stored under the desk. More importantly, however, we found that the heaviest users of their smartphones—those who were younger and grew up with technology—showed increased anxiety after just ten minutes of not being able to use their phone, and their anxiety continued to increase across the hour as compared to those who used their phones less. This rise in anxiety is also evident in the prevalence of phantom pocket vibration syndrome.[52] Given research showing how often this occurs and how often younger, more technologically immersed people check their smartphones, it is hardly surprising that in the study described earlier, technology use was found to be related to symptoms of psychiatric disorders.

In the next chapter, we will expand on the mental health ramifications of the interfering nature of technology in individuals with clinical conditions, but first let's examine how our Distracted Mind can negatively affect our sleep, an important determiner of our mental and physical health.

Sleep

Before more closely examining how lack of sleep affects people, let's examine what happens during sleep and what mechanisms might be responsible. Myriad evidence indicates that the process of sleep, in well-rested people, is fairly patterned. During the day, sunlight increases exposure to more short blue wavelength light, which increases our alertness through the release of cortisol. As daylight begins to fade, giving way to dusk, red wavelength light predominates, which allows melatonin to increase, which then helps us fall asleep. The process of melatonin release is a slow one, beginning about two to three hours before bedtime. Once we fall asleep, our brain activity goes through four phases from light sleep to deep sleep, which is then followed by rapid eye movement (REM) sleep that signifies that we are dreaming. In a normal night's sleep, this process repeats four times, with REM sleep getting longer as the night wears on. During sleep the brain performs a variety of housekeeping chores called "synaptic rejuvenation" including pruning and memory consolidation, which serve to remove unimportant connections and enhance important ones. In addition, our nighttime brain flushes out toxins that are by-products of daytime neural activity, which if not removed can have deleterious effects on neurons in the brain.

Photopigment cells in the retina at the back of your eye help control the release of melatonin. In order to produce white light, technology screens must emit light at multiple wavelengths, including blue short wavelengths. When you are exposed to blue light, the photopigment cells signal your brain that it is a time to be alert. Given our children's and our own predilection to use our technology in the bedroom right before sleep, we are bombarding our eyes with blue light that signals awake time rather than red light that signals it is time to go to sleep. In fact, blue light is far stronger when one is looking at a small screen held close to the face—as most people do with smartphones and tablets—than for larger screens such as televisions, which are viewed farther away.[53]

Work by the National Sleep Foundation (NSF) as well as a recent study of 362 adolescents and a meta-analysis of sixty-seven studies of the impact of screen time on children and adolescents found that screen time, particularly in the last hour prior to sleep, is related to sleep problems, primarily resulting in fewer nightly hours of sleep and poorer sleep quality.[54] In addition, studies have shown that 47 percent of college students awaken at night to answer text messages, and 40 percent awaken to answer phone calls, resulting in forty-six minutes less nightly sleep.[55] With the vast majority of teens using a variety of technologies prior to sleep as well as awakening during the night to address smartphone alerts, it is likely that their brains are not getting the nightly housekeeping that was mentioned above, which can lead to mental difficulties.

Research has shown that over the past fifty years there has been a decline in the number of hours of nightly sleep as well as in the quality of that sleep, with one factor being attributed to the statistic that 90 percent of American adults use their electronic devices within an hour of bedtime at least a few nights a week.[56] A study of more than 2,000 fourth- and seventh-graders found that children who slept near a small screen device had nearly twenty-one minutes less sleep than those children who did not sleep in close proximity to a phone or tablet, and those who slept in a room with a television set reported eighteen fewer nightly minutes of sleep.[57] Total minutes of nightly bedroom screen interactions were strongly related to sleep problems.

A recent study by researchers at Harvard Medical School examined the effect of reading from an e-book compared to reading from a paper book on nighttime sleep and morning alertness and found exactly the results we might expect.[58] Compared to reading from a paper book, reading from an e-book led to an average of ten minutes longer to fall asleep, delayed melatonin onset by an hour and a half, reduced melatonin release an average of 55 percent, reduced valuable rapid eye movement sleep by twelve minutes, and reduced morning alertness. Granted this was a study of only twelve college students, but they participated for two weeks as inpatients and had their blood taken hourly as they read a print or e-book for four hours prior to attempting to fall asleep and as they slept.

A recently published study by Dr. Rosen's lab examined sleep and nighttime technology use habits among 391 American college students ranging in age from eighteen to sixty-nine and found that only 19 percent of them put their phone away or on silent when they went to sleep; 81 percent kept their phone close by on vibrate (39 percent) or with the ringer left on (42 percent).[59] Overall, half the students checked their phone when they awoke at night, which matched a previous study.[60] Using a series of measurement tools to assess sleep quality, executive functioning, anxiety about missing out on electronic communications, and daily smartphone use, we discovered four paths to a predicted poor night's sleep. Poorer executive functioning (ability to make good decisions) predicted both more smartphone use, and poorer sleep quality and anxiety about missing out predicted more smartphone use and more nighttime awakenings both leading to poorer sleep. This means that both our ability to make smart nighttime choices as well as our anxiety about what we might miss out on in our virtual worlds during our sleep combine to disrupt our sleep, which then leads to poorer thinking skills and more nighttime interruptions. This is indeed a downward spiral that results in disrupted mental functioning.

Research has also demonstrated that getting too little sleep can disturb memory in important ways. Of course, memory is a complex process that involves multiple areas of the brain, but one key component to memory is a solid cognitive control system without which the information would never get transmitted effectively or completely to memory centers such as the hippocampus. One study found that adults who routinely slept less than five hours a night were more likely to incorporate misinformation in their morning report of either photos or videos that they had observed before bedtime; some even reported that they had seen video footage of an event that never happened.[61] Another study kept college students awake for twenty-four hours—not necessarily an unusual occurrence in their lives—and had them view a videotape showing a man stealing a woman's wallet and putting it in his jacket pocket. Forty minutes later, they were told that the man put the wallet in his pants pocket. The sleep-deprived students were more likely to concur with this misreported location of where the wallet was stowed than those who slept a full eight hours.[62] Other studies have corroborated these memory decrements, and the American Academy of Pediatrics is now urging schools to start later to avoid these memory and cognitive control deficits faced by sleep-deprived students.[63] A recent Mayo Clinic study that examined the amount of light that suppresses melatonin recommends that if you want to use your device in a darkened bedroom, dimming the smartphone or tablet brightness settings and holding the device at least fourteen inches from your face will not suppress melatonin and thus will not interfere with your sleep.[64]

Although most sleep studies have been performed with children, teens, and young adult college students, some research has examined the impact of sleep duration and quality for workplace adults. For example, one study of US managers and employees found that those who used their smartphones for work at night after 9 p.m. showed impaired cognitive control at work the following day, which was reflected in reduced attention to work and increased depletion of working memory resources.[65] A similar study of Belgian adults found that having the Internet in the bedroom resulted in similar negative effects on cognitive control.[66] In chapter 11, we will discuss nighttime routine changes that will enhance our sleep by allowing our brain to do an effective job of nightly housekeeping and help us become more functional during the day.

In the previous two chapters, we showed how technology is a major aggravator of the Distracted Mind and elucidated its consequences ranging from harming workplace and school performance, damaging relationships, and eroding our safety, sleep, and mental health. In the next chapter, we will explore how technology affects the Distracted Minds of children, adolescents, older adults, and individuals suffering with mental health conditions.

8 THE IMPACT OF TECHNOLOGY ON DIVERSE POPULATIONS

THE RAPID RISE of information technology has changed the way we interact with the world around us. We documented how the influx of high-tech into every aspect of our lives can be traced to three game changers: the Internet, smartphones, and social media. And we showed how carrying a computer in our pocket or purse has offered us constant access to a broad social and informational world. The empirical research presented in the last two chapters highlights some of the negative aspects of the impact of this technology on the lives of healthy adult populations, including students and office workers. In this chapter, we extend this discussion to populations who are uniquely affected by technology. We explore how individuals with greater baseline cognitive control limitations manage the multiple streams of information that engender goal interference. In other words, how does technology affect the behavior of various populations in the context of challenges that they already face with their more Distracted Minds?

We begin by examining how children and teens are more prone to interference and then extend this discussion to tackle how older adults' daily activities—sleeping, walking, and driving—are prone to interference from technology, which then can have major consequences on successful functioning. Next we explore how people with psychological impairments deal with technology-based interference, ranging from symptomology found in depression and anxiety to attention deficit hyperactivity disorder (ADHD), and even narcissistic personality disorder, to those suffering from autism spectrum disorders.

Dr. Rosen has studied and written extensively about the impact of technology on five generations of Americans—baby boomers (born between 1946 and 1964), Generation X (1965–1979), the Net Generation (1980–1989), the iGeneration (1990–1999), and a newly labeled "Generation C" born in the new millennium.[1] In this research and that of our colleagues, we have shown that although the younger generations believe that they can multitask better than older generations, their real-world behaviors and performance suffer when they concurrently engage in activities using multiple forms of media.[2] These performance decrements, which occur despite superior cognitive control abilities in this age group, may be exacerbated by the fact that children, teens, and young adults have grown up in a world where technology has entered and permeated their lives more rapidly than for any previous generation. While baby boomers grew up in a world where technologies such as the telephone, television, and computer took as long as ten to twenty or more years to penetrate society—giving them time to evaluate, learn, and assimilate the new media—their children and grandchildren have seen technologies such as the iPhone and iPad, websites such as Facebook, MySpace, Instagram, and Pinterest, and games such as Angry Birds and Words with Friends, enter and invade their world in a matter of only a few months. Early technology adopters, sometimes called "Digital Natives,"[3] are members of the Net Generation, the iGeneration, and Generation C who eagerly embrace attention-grabbing technologies without taking the time to recognize and appreciate how they might best engage with them, in terms of the affect on their brains and lives.

We learned in earlier chapters that the prefrontal cortex, which is critical for decision making, goal setting, and all cognitive control functions, does not completely mature until a child reaches young adulthood. We also know that the cortex does not complete its development, particularly in the area of the "social brain" (defined by neuroscientist Sarah-Jayne Blakemore and her colleagues at University College London as including the medial prefrontal cortex, temporal–parietal junction, posterior superior temporal sulcus, and anterior temporal cortex), until young adulthood.[4] Myelination—the process whereby neurons are coated in fatty cells to allow faster and more accurate transmission of signals between neurons and brain regions—also does not conclude until young adulthood.[5] Based on these aspects of normal development, and research on ramifications of early television watching on later attention problems, the American Academy of Pediatrics recommended that "television and other entertainment media should be avoided for infants and children under age 2 and used minimally for children over the age of 2."[6] In spite of that recommendation, recent studies have shown that most children use the Internet daily. In one study of children under age eleven in the United States, 21 percent of boys and girls used the Internet once a day, and an additional 39 percent used it many times a day.[7] Studies from other researchers have shown similar trends, with children using digital media for an average of 5.5 hours per day with the predominant media platform being television (3.5 hours per day), and a shift to Internet-based media beginning at around eight years old.[8] Other studies have reported that those 5.5 hours are, in reality, closer to eight hours when you accept the fact that children are already attempting to multitask while consuming media, by combining two media forms such as the television and an iPad, by combining the use of one media form and a real-life situation such as being at the dinner table, or by performing multiple tasks on one media platform (e.g., a computer or iPad).[9]

What is the potential impact on children of consuming massive amounts of media from a young age? Aren't educators striving to put computers and tablets into all levels of the educational system? Doesn't that mean they feel that the good outweighs the potential problems? According to two recent studies, three out of four K–12 teachers asserted that student use of entertainment media (including communication tools such as social media) has hurt students' attention spans a lot or somewhat, 87 percent of teachers reported that the use of technologies is creating "an easily distracted generation with short attention spans," and 64 percent felt that digital media "do more to distract students than to help them academically."[10] That, in itself, is not a promising assessment, but there is more. Most people are amazed at how easily very young children take to a touchscreen environment, and parents often report that their children just seem to know intuitively how to scroll through an iPad or an iPhone to find the video they want to watch or the game they want to play. According to the Joan Ganz Cooney Center, nearly two of every three children between the ages of two and ten have access to a tablet or an e-reader.[11] A recent article in *Wired* that summarizes research on touchscreen use by children includes the following quote from the author in his role as a parent:

> But these screens have a weird dual nature: They make us more connected and more isolated at the same time. When I hand my daughter an iPad with an interactive reading app, she dives in and reads along. But she also goes into a trance. It's disturbing because, frankly, it reminds me of myself. I'm perpetually distracted, staring into my hand, ignoring the people around me. Hit Refresh and get a reward, monkey. Feed the media and it will nourish you with @replies and Likes until you're hungry and bleary and up way too late alone in bed, locked in the feedback loop. What will my daughter's loop look like? I'm afraid to find out.[12]

As families and children have embraced these devices, we are starting to discover the impact that these technologically immersive environments have on young minds. Dr. Rosen's lab examined the relationship between media use among children, preteens, and teens and "ill-being," which included physical health, psychological issues, behavioral problems, and attention difficulties.[13] Even after factoring out the effects of parent and child demographics—socioeconomic status, parent and child BMI, and other characteristics that might influence health—as well as factoring out eating habits and daily exercise (all shown to be predictive of health), the total amount of daily media consumed by children predicted ill-being in all four categories. The more media children, preteens, and teens consumed daily, the less healthy they were. In addition, preteens and teens were less healthy as they played more video games regardless of total media consumption. Spending more time on the Internet also increased ill-being for teenagers. Other researchers have shown similar negative health effects for children in general psychological difficulties, poor psychological adjustment, emotional and family problems, social well-being, impulsivity and attention, obesity, family functioning and friendships, and academic performance.[14] Study after study has indicated that *too much* technology use, whether it be watching television, going on the Internet, using a smartphone or tablet, or playing video games, is associated with deleterious effects on the health of our children.

Of course, as with all correlational data there is the caveat of the directionality of these effects. It may be the case that children with these issues seek out the high-tech world more readily. In either case, caution on excessive use seems warranted. Although we have passed beyond the state where we can keep all media away from children under two (seriously, how can you not allow your eighteen-month old to watch videos or play games on your iPad when it is so effective in distracting them and making your day easier?), we can exercise sound behavioral principles to regulate those behaviors and create an environment where the interfering nature of technology is lessened and some of its use is replaced by different activities. We will discuss our ideas on how to make this work with a minimum of tears in chapter 11.

One issue that cuts across all populations, but particularly children, adolescents, and young adults, is the impact of technology on sleep. The National Sleep Foundation performs annual national sleep surveys, and in the most recently published report in 2014 NSF surveyed 1,103 adult parents of six- to seventeen-year-olds and confirmed what all parents know: children and teens are decidedly short on sleep and most likely running up a massive sleep debt.[15] The statistics are somewhat staggering. The accepted required amount of sleep for children and teens is nine hours per night. In the NSF study only 10 percent of teens got their nine hours on weeknights, compared to 19 percent of preteens and 69 percent of children. Strikingly, 56 percent of teens and 29 percent of preteens slept less than seven hours. At that rate, a five-day school week would generate a massive sleep debt in the majority of teens. Parents most certainly realize this when their teens and preteens sleep in on the weekends in an attempt to pay off that debt.

The NSF study also found some disturbing statistics on the use of technology in the bedroom and its impact on sleep. Across all six- to seventeen-year-olds, at bedtime 45 percent had a television in the bedroom, 40 percent had a music player, 30 percent had a tablet or smartphone, 25 percent had a video game, and 21 percent had a computer with one in three bedroom televisions left on at night after the child or teen fell asleep. In fact, only one in four children or teens did not have a single device in their bedroom at bedtime, and 97 percent of adolescents had at least one bedroom device.[16] Those who had *any* electronic device in their bedroom showed an average of forty-two minutes less sleep per night. And having a smartphone in the bedroom at bedtime reduced their nighttime sleep an average of fifty-four minutes. Lest we believe that this is an American phenomenon, research in Egypt, New Zealand, and Finland found similar patterns of behaviors by children and adolescents.[17]

Teenagers face similar issues to younger children, but the ramifications are more serious. Adolescents are constantly faced with decisions that can have severe repercussions, including physical harm. Faced with the similar brain maturation issues as children, the older adolescent is making the transition from school to either the workforce or additional schooling, is able to drive, can hold down a job, and can enter into relationships that are laden with emotions. Situations inviting the adolescent to engage in potentially risky behaviors are commonplace.

Adolescents are simply not always equipped to handle complex decisions that require careful thought and planning because of their underdeveloped prefrontal cortex as well as incomplete connections between brain areas that are critical in assessing situations. For example, research has demonstrated a relationship between developmental delays in the prefrontal cortex and the inability of adolescents to control their impulses, inhibit poor responses, and avoid the "hyper-responsive" reward system that deals with emotions and drives behavior in spite of potential negative consequences.[18] In addition, a wealth of data suggests that because of the developmental process, adolescents are less capable of assessing nonverbal cues, which then directly hinders their ability to be empathic toward others and have effective communication with peers and families.[19]

Adolescents are rarely out of sight of at least one technological device— and most often two, three, four devices or more—insistently competing for their attention and causing external interference. Even when devices are out of sight, young minds are still thinking about what might be occurring in virtual worlds, coaxing them to get back to the smartphone, tablet, or laptop to "check in" (creating internal interference). And attempts to live without these devices by going on a technological detox or fast simply won't work, since they cannot be separated from technology without feeling lost or out of touch.[20]

Here are interesting quotes from two typical adolescents involved in a Kaiser Family Foundation study defining their relationships with technology:[21]

> I multitask every single second I am online. At this very moment, I am watching TV, checking my email every 2 minutes, reading a newsgroup about who shot JFK, burning some music to a CD and writing this message. (seventeen-year-old boy)

> I'm always talking to people through instant messenger and then I'll be checking email or doing homework or playing games AND talking on the phone at the same time. (fifteen-year-old girl)

The bottom line is that our youth are constantly immersed in a high-interference environment whose strong bottom-up factors compete for their attention; and when these external enticements capture attention, then top-down, goal-driven behavior face severe challenges.

OLDER ADULTS

Although technology is often considered a tool for the young, they are not the only ones engaging in the high-tech world. Another study from Dr. Rosen's lab compared five generations of Americans on their technology use.[22] Interestingly, although baby boomer adults showed the least technology usage of all generations, a substantial proportion of them used technology regularly. We asked individuals from all generations how many hours per day they used various technologies with the understanding that many of those reports were inflated by overlapping use. The most technologically active groups—iGeneration teens and Net Generation young adults—used technology about twenty hours per day, again exaggerated by multitasking. As expected, older adults showed fewer daily hours of use but their use still remained high at nearly 12.5 hours per day. When looking at which technologies baby boomers preferred, the most popular were: watching television (2.4 hours per day), talking on the phone (1.9), doing offline computer tasks (1.6), sending and receiving email (1.5), and listening to music (1.5). In addition, although older adults have been viewed as "digital immigrants," 71 percent of them regularly sent and received an average of 214 text messages per month, or about seven per day. They may not reach the teen ranks with their reported 3,417 texts per month, but the data still show that older adults are using newer technologies that were thought to be only the province of digital natives.[23]

Other research confirms and expands on the idea that older adults are using technology regularly. A recent Pew Research Center report compared seniors with all other American adults and found that many older adults have "relatively substantial technology assets, and also a positive view toward the benefits of online platforms."[24] In comparing a similar nationwide study done just a year earlier, Pew found that 59 percent of seniors report that they go online, representing a 6 percent increase in just one year. Strikingly, 71 percent of Internet-using seniors go online daily while another 11 percent go online three to five times a week. Another study found that older adults attempted to multitask with two or more forms of media for more than 90 minutes a day.[25]

Despite the increase in use, there are emerging data on the beliefs and attitudes of older adults toward technology that suggest hesitancy in adoption. The study from Dr. Rosen's lab that queried five generations about their comfort in using technology to reveal that baby boomers demonstrated more anxiety than any other generation. That anxiety appeared to be a combination of negative attitudes about the value of technology but also about their confidence in understanding and using technology.[26] Other studies have confirmed the role that anxiety plays in older adult technology use.[27]

However, it would appear that this anxiety can be overcome with exposure. A review of more than 150 studies involving computerized cognitive training with older adults concluded that "despite common misperceptions [that] older adults do not enjoy learning to use new technology, perceptions of the computerized training programs were positive for the older adults who completed computerized training. In spite of many older adults reporting anxiety about using unfamiliar technology at the beginning of training, most reported high levels of satisfaction after training was completed."[28]

In earlier chapters, we discussed how older adults suffer from diminished cognitive control across all areas of attention, working memory, and goal management. Research from the Gazzaley Lab demonstrated that when it comes to distraction, reduced suppression abilities conspire to derail them by failing to filter irrelevant information in their environment. It seems likely that technology may aggravate this distraction effect. A recent study examined the effects of background music on the ability of adults of all ages to perform a simple task of matching faces. While all adults, young and older, rated the background music as distracting, only the older adults did not perform as well with background music as they did with silence.[29] This is suggestive of challenges older adults may have in the modern open-plan office workplace.

When it comes to multitasking, the Gazzaley Lab showed that older adults fail to reengage their prefrontal cortex networks after an interruption, which means that they are still internally focused on the interruptor even though its external presence is gone. Again, this suggests a potential area of concern for the involvement of high-tech. One study compared young adults with older adults as they used a smartphone to play a simple game while walking. Interestingly, while younger and older adults showed decrements in both walking and game performance, older walkers showed poorer game performance and a more irregular gait as they walked and played the game. Older adults also showed different prefrontal cortical areas mediating their performance perhaps as a result of demyelination.[30] Another study that observed a range of younger to older adults found that the older adults were the most negatively impacted crossing an intersection when distracted.[31] Given earlier studies showing the dangers of multitasking and walking, coupled with increased smartphone usage by older adults, it is important to recognize that they may face more potential safety hazards when walking and using technology.

Driving an automobile is an even more serious undertaking that requires cognitive control processes. According to a recent study that examined the driving and technology use habits of older adults, nearly all use some form of technology when driving, with the radio being used by 91 percent of elderly drivers followed by eight in ten adjusting electronic dashboard controls and half using a cellular phone.[32] From the research that we have presented thus far, the use of these devices while driving poses an increased risk of accidents among all drivers but most certainly among older adult drivers. A recent publication by the World Health Organization examined the research on driver distraction and concluded that while all drivers are impaired by their use of technology while behind the wheel, older adults face special problems because of their decreased visual and cognitive capacities, which make it even more difficult for them to combine driving with any other task.[33]

CLINICAL CONDITIONS

The symptomatology of many psychiatric and neurological disorders is characterized by real-world behaviors that indicate a lack of cognitive control. As described in chapter 5, individuals with psychiatric conditions such as depression, schizophrenia, and ADHD and neurological disorders such as Alzheimer's disease have greater difficulty ignoring irrelevant information, attending to relevant stimuli, inhibiting responses, selecting relevant information from memory to apply to a current situation, engaging working memory, and switching flexibly between multiple goals. This can often be traced to problems in the prefrontal cortex and its networks with the rest of the brain that mediate cognitive control.[34] The question we will address here is how modern technology affects the behavior of people with psychiatric and neurological disorders in the context of their Distracted Minds. In the next sections of this chapter we will examine research on several disorders: ADHD, depression, anxiety, narcissistic personality disorder, and autism.

ADHD

Attention deficit hyperactivity disorder, as well as ADD without hyperactivity, is on the rise among five- to eleven-year-olds, with estimates indicating an increase of 25 percent over the past decade.[35] A recent meta-analytic study, which examined forty-five empirical published research reports on the impact of media use on ADHD among children and adolescents, found significant relationships between the use of television, video games, and general media of either a violent or nonviolent nature and ADHD-related behaviors.[36] Other work suggests that this impact may be related to less activity in brain areas that are associated with attention including the frontal, parietal, and temporal cortexes, the striatum, and the cerebellum.[37]

Given the literature suggesting that children and teens multitask more than any other group, those with ADHD are facing what the researchers have termed a "bottleneck," in which executive functions and cognitive control are stalled.[38] Commenting in a blog for the Dana Foundation—a philanthropic organization that supports brain research—Johns Hopkins professor Martha Denckla says, "Multitasking to me fits into the realm of attention and executive functions. You can measure a timed response where they have to shift or multitask, and you see that ADHD kids are slower. We all have a central processing bottleneck ... it costs everybody something ... you have a greater refractory period, but they have more."[39] Using a tool called "Six Elements Test"—which has research participants devise a plan, schedule work, and keep track of time—children, preteens, and teens with ADHD attempted fewer tasks, showed poorer planning, demonstrated worse goal management, and exhibited reduced working memory than normal cohorts, all integral parts of cognitive control that are required for peak intellectual performance.[40] This suggests that the high-tech of today may exacerbate preexisting cognitive control deficits and negatively influence real-world performance in young people who are suffering from this condition.

A recent study by researchers at the Cincinnati Children's Hospital examined cognitive processing decrements in adolescents with and without ADHD as they drove in a driving simulator under three conditions: talking on a cell phone, texting on a cell phone, and a control condition where they drove without additional potentially distracting tasks.[41] Regardless of whether they had a diagnosis of ADHD, adolescents who texted while driving exhibited signs of driving impairment: they drove more slowly, their speed was more variable, and their position in the lane was more variable than when they drove with no additional tasks. Talking on the cell phone served to reduce the variability of their lane position but did not affect speed. However, when driving in a simulator, under all conditions, even without the use of additional tech, those adolescents diagnosed with ADHD showed both more variability in their driving speed and their lane position. This interesting study offers insights into the negative impact of technology on adolescent drivers in the setting of their additional burden of having a baseline attention deficit, or as the authors refer to it, "the combined risks of adolescence, ADHD, and distracted driving."

Depression and Anxiety

In a 2011 report, the American Academy of Pediatrics Council on Communications and Media declared that social media might be a potential cause of "Facebook Depression" that, they claimed, develops when preteens and adolescents spend excessive time using social media.[42] Since then, there has been a rash of research examining the mechanisms by which media might induce new symptoms of depression or increase preexisting disposition toward depression. The results have been mixed. Earlier we mentioned research where Dr. Rosen's lab found that a combination of technology use and preference for multitasking predicted symptoms of nine psychiatric disorders including both mood disorders and anxiety-based disorders. Interestingly, while seven of the nine tested disorders showed an association between technology use (mostly social media use) and increased symptomology, symptoms of two disorders—dysthymia (mild depression) and major depression—actually showed the opposite pattern. Those teens and adults who had more Facebook friends *and* who talked on the phone more often showed less depression. Although this was only a correlational study (albeit while factoring out many alternative explanations), it suggests that having more Facebook friends and being able to talk to them directly on the telephone may be helpful in ameliorating depressive symptoms.[43] In concordance with this study, a longitudinal assessment of Australian adolescents found that over a twelve-month period, increased use of the Internet for social reasons (including instant messaging and social media) predicted less depression and less compulsive Internet use, which was related to how satisfied the adolescents felt with their social support, their self-esteem, and their coping strategies, which the authors assert are part of the youths' abilities to moderate their cognitive control abilities.[44] Other studies with college students have found similar results.[45] In terms of the causal directionality of these conclusions, the impact of social media posts on symptoms of mood disorders was informed by one study that actually directed college students to post more than usual on Facebook for one week. At the end of that short time period, the students reported feeling less lonely as a result of feeling more connected to others.[46]

Some studies have found a clear negative association between Facebook and other technology use with mood—particularly violent video game playing—among preteens, high school students, and young college-age adults in the United States and around the world.[47] However, in a meta-analysis of eighteen studies, a team of researchers from the University of Wisconsin-Milwaukee concluded that the link between Facebook use and mood is not that Facebook causes loneliness but that it is more likely that people tend to use Facebook more often when they are lonely.[48] Given that depression has been shown to be related to cognitive control issues such as rumination—repetitive focusing on unpleasant stimuli—or disrupted executive functioning in general, it makes sense that those who have symptoms of depression and are ruminating might seek help through social media.[49]

Researchers at Missouri University of Science and Technology assessed depressive symptoms among more than 200 college students and then monitored their Internet activity on the campus servers for a month.[50] Results indicated that those who were more depressed at the outset of the month-long study showed more frequent switching between computer applications, indicating that they have more difficulty managing their goals. Finally, in a recent controversial study, researchers manipulated the items that Facebook users would see on their wall to include either more positive posts or more negative posts. They found that after viewing more positive posts users would post more positive messages themselves.[51] The reverse was true for negative posts, suggesting that emotional content encountered online is "contagious" and spreads from person to person as they attend to messages that then promote similar internal feelings.

Finally, a study by researchers at Michigan State University gave a complete battery of online surveys to 318 university students that measured media multitasking and various aspects of psychosocial dysfunction.[52] Even after controlling for how much media the student used each day as well as personality traits such as neuroticism and extraversion, both of which have been shown to be related to depression, they found that students who engaged in more media multitasking showed more symptoms of depression as well as more symptoms of social anxiety. In fact, media use in general was not associated with social anxiety, but media multitasking, attempting to use more than one media form at the same time, was related to social anxiety.

In the previously mentioned study that examined predictors of symptoms of psychiatric disorders, all anxiety-based disorders were associated with some combination of media and technology use, including time spent online, social media, and electronic communications.[53] Although research on the impact of technology on anxiety-based disorders is scant, we do know that there is an inverse relationship, such that anxiety disrupts cognitive control, particularly diminishing working memory and disrupting attention and goal management.[54] In chapter 11, we will discuss ways to reduce anxiety and, thus, increase cognitive control to reduce internal interference.

Narcissistic Personality Disorder

Researchers Jean Twenge and Keith Campbell have presented evidence that over the past two decades the level of narcissism in college students has increased dramatically.[55] In the study out of Dr. Rosen's lab mentioned earlier in this chapter with respect to symptoms of depression, three different uses of technology predicted increased symptoms of narcissistic personality disorder: having more Facebook friends (the opposite result found for depressive symptoms), using Facebook more on a daily basis, and using Facebook more for managing one's impression including posting comments and/or posting pictures of oneself.[56] These Facebook behaviors may reflect an individual's baseline difficulties with cognitive control, including an inability to attend to material that is not related directly to him or her, and suggests that excessive social media may exacerbate symptoms of narcissism.[57]

Autism Spectrum Disorders

Autism has as one of its characteristics a rigid focus of attention on objects or details and difficulty switching focus—possibly reflecting pathology in the striatum or frontal lobes—clearly indicating the presence of a cognitive control deficit.[58] An experiment using the "virtual errands task," where college students with autism spectrum disorder (ASD) had to perform a series of planned errands in a virtual university, indicated that these individuals suffer from executive problems including planning inflexibility, inhibition difficulty, and an inability to recall task requirements.[59] In addition, researchers have reported finding social cognition or "theory of mind (ToM)" deficits in adults with Asperger's syndrome, a form of autism.[60] In this study, individuals with Asperger's syndrome were found to have difficulty identifying the emotion a person was feeling when shown images of a partial face from midway along the nose to just above the eyebrows. One argument that has been raised as to why people with ASD might not be as good at assessing emotions or thoughts of others is that they spend too much time using technology in lieu of interacting face to face with others, although once again, causal conclusions cannot be reached from these data.[61]

In looking at technology use among those with ASD, one possibility is that individuals with more severe symptoms or lower intellectual functioning may spend more time using less cognitively demanding media, such as television; while those with milder symptom presentations may have increased engagement in a wider range of activities. Alternatively, excessive use of some screen-based media may be associated with the presence and intensity of restricted interests for children with ASD, irrespective of overall severity. Prospective longitudinal designs would be helpful in determining whether and how screen-based media use influences functional outcomes and developmental trajectories of children with ASD. It will be particularly important to determine whether such effects are significant even after controlling for initial levels of severity and functioning.

In this chapter, we explored the impact of technology on goal interference in populations with cognitive control deficits, which supported the general conclusion that the use of modern technology exacerbates the preexisting challenges these individuals face in effectively interacting with their environment. In the next chapter, we will examine the potential underlying causes for why almost all of us seem to interact with our technology in such a frenzied multitasking manner, so that we stress our fragile cognitive control system and aggravate the collision with our goals.

9 WHY DO WE INTERRUPT OURSELVES?

What information consumes is rather obvious: It consumes the attention of its recipients. Hence a wealth of information creates a poverty of attention, and a need to allocate that attention efficiently among the overabundance of information sources that might consume it.

—NOBEL PRIZE–WINNING ECONOMIST HERBERT SIMON[1]

WE HAVE DESCRIBED three "game changers" in our rapidly changing high-tech world: the Internet, smartphones, and social media. Each features, at its core, a commodity that is precious us—information. These technologies have the remarkable ability to grab our attention at any instant with an alert, notification, buzz, beep, or even simply in response to a thought that it must be time to check in and make sure that we have not missed out on anything important in the past few minutes. What just a few short years ago felt like precious gifts of connectivity to vast stores of information, as well as access to people—providing even more information—has now started to feel like a burden. None of this should be news to us. In fact, in a *New York Magazine* article back in 2009, Sam Anderson warned us that "the virtual horse has already left the digital barn. It's too late to just retreat to a quieter time. Our jobs depend on connectivity. Our pleasure-cycles—no trivial matter—are increasingly tied to it. Information rains down faster and thicker every day, and there are plenty of non-moronic reasons for it to do so. The question, now, is how successfully we can adapt."[2] And adapt we must. Unlike the squirrel who knows instinctively and reflexively when it is time to stay put and continue to eat the available acorns and when it is time to move to a new patch of acorns, we must learn to make conscious decisions about when it makes more sense to stay attentive to what we are doing and avoid those multisensory calls for attention.

To accomplish our ultimate goal of gaining control, the first step is to understand why we are so driven to engage in using new technology with such an extreme multitasking style that it places pressure on our cognitive control abilities and results in so many negative life consequences. This behavior is such a salient aspect of our modern lives that it has come to be viewed as the epitome of the Distracted Mind. In chapter 1, we proposed a hypothesis that we engage so frequently in interference-inducing behaviors—such as media multitasking in noisy places—because from an evolutionary perspective we are merely acting in an optimal manner to satisfy our innate drive to seek information. As described, we are inherently information-seeking creatures who forage for information resources in a manner similar to how our ancient ancestors foraged for food. In this chapter, we will assess how several aspects of modern technology fuel this innate drive, and further consider the possibility that conditions are such that we are no longer behaving optimally, even when considered from a purely information-foraging perspective.

In chapter 1, we introduced the marginal value theorem (MVT), which has been used for decades to explain why, how, and when animals take the time and energy to travel to a new patch with additional food, rather than gather dwindling food resources from their current patch. In its simplest form, the MVT explains the cost–benefit relationship of remaining in a food patch versus moving to a new patch, with an animal's drive to survive as the instinctive force to accumulate resources. An animal's ability to find nourishment by spending an "optimal time" at one source before traveling to another is a critical factor in its survival. To review the MVT model, in figure 9.1 the right side of the model indicates the *benefits* of remaining engaged in a resource source. For animals foraging for food, the *resource intake curve*, which defines the cumulative resource intake over time, is driven largely by external factors that define the diminishing gains of remaining in a patch, such as the number of nuts left on the tree. Or more precisely, how much time has passed since the squirrel found that last nut? As resources dwindle with continued consumption, the benefits of remaining in a given patch decrease and the curve gradually flattens over time. This plateauing of the resource intake curve on the right side of the model interacts with factors on the left side of the model, which reflect the *cost*, or the time to get to a new food source (how long will it take to get to a new tree full of nuts?). It is the intersection of these two factors (expressed by the tangent line to the intake curve) that is associated with a subconscious trigger in an animal's brain that signals the *optimal time to remain in a source*. For animals foraging for food, instinctive forces drive this behavior.

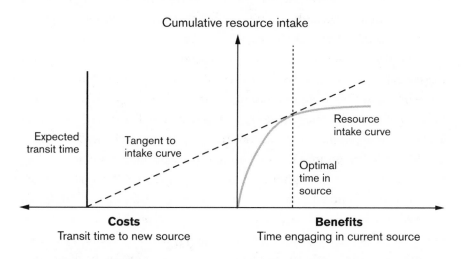

Figure 9.1
A graphical representation of the marginal value theorem.

Let's now consider the MVT model in terms of our interactions with technology, notably to illustrate how it can help explain our seemingly constant need to switch our focused attention in response to either an external alert or an internal trigger. Does our short attention span map onto the squirrel's transition from patch to patch and its instinctive drive to gather resources to survive? Clearly much of what we are doing with our technology is not related to survival, and yet we appreciate that we forage for information in a manner similar to other animals foraging for food. Study after study, both in the laboratory and in the field, has demonstrated that adults, teenagers, and even children are shifting their attention to a new information patch even before they have completed their task in the original source. And we have seen how this wastes time in the form of a "resumption lag" or the period it takes to return to the original task and recall where we left off, or as we often must do, retrace our steps to once again engage with material on which we should be focusing. We have also shown how frequent attention shifts result in stress as we attempt to perform a task in a limited amount of time. However, we don't face the same threat that the squirrel faces of having to stay alive and safe, all the while keeping an eye on both the dwindling resources in the current location and the potential for more resources in another patch. So, does this model offer any insights to our behavior?

The MVT model has recently been expanded from explaining why animals move from one food source to another, to explaining why humans move from one source of information to another.[3] We propose that a human information foraging model, and specifically the MVT, can also be used to explain why we media multitask so avariciously. It may explain, for example, why we may choose to: (1) stop working on a document we are reading online to check our smartphone for an incoming alert, (2) open a new tab to search for additional information on an unrelated topic, and (3) decide we need to text a friend to arrange an evening out, *all* before returning to our document. And then we are faced with having to remember where we were and rebuild the mental representation of the material in the document.

In considering how the MVT model might explain how modern technology has influenced our Distracted Mind, the main question we need to address is why so many of us frequently choose to move rapidly from one high-tech media activity (information patch) to another. Along the way we will consider whether the MVT model explains the factors that influence this behavior, and if, because of modern technology, we are no longer performing "optimally" in accomplishing our goal of foraging for information. In chapter 11, we use this same model to guide the formulation of strategies to help us spend a more optimal time in an information source, with the ultimate goal of resolving our interference dilemma and improving the quality of our lives.

TECHNOLOGY INFLUENCES

Humans foraging for information clearly face external factors that influence the resource intake curve in a manner similar to what occurs for a squirrel faced with dwindling nuts in a tree. Let's say that you sit down to respond to the day's email accumulation and find that you have twenty unanswered messages to wade through. After thirty minutes spent in your inbox information patch, you have only a couple of emails left. At this point, the benefits of continuing to check your email are less than when you first sat down. This diminishing benefit of remaining in an information source is reflected over time as a flattening resource intake curve. Many other information patches display similar dynamics. For example, when engaging in a back-and-forth text exchange, there often comes a time when the value of information being shared starts to approach zero (let's be honest here, we know this is the case when the last few texts or emails are single words or emojis). In this manner, external factors result in the information resource intake curve flattening more rapidly with the passage of time, just as it does for food patches, and this drives the information forager to shift to a new patch, especially if the expected transit time to a new patch is a short hop to another webpage or another app on our omnipresent smartphone screen.

But interestingly, in humans it seems that the situation is much more complicated than the influence of external factors on the resource intake curve. We propose that the right side of the model is also strongly influenced by *internal factors* that modulate the slope of the curve, independent of actual diminishing information resources (the external factors). It seems that at least two internal factors exert a major influence in flattening the curve while we actively engage in information foraging: *boredom* and *anxiety* (seen on the right side of figure 9.2). We hypothesize that the accumulation of these two internal signals with the passage of time while engaged in an information patch—reflected physiologically as diminished arousal and increased stress, respectively—lowers the peak of this curve. This, in turn, drives the "optimal time in source" to the left (i.e., reduced time in source), thus leading to more rapid switching between information patches.

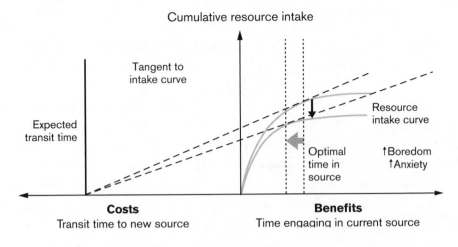

Figure 9.2

Increased boredom/anxiety have flattened the slope of the resource intake curve to result in diminished time at an information source, and thus more rapid task switching.

As we will describe in this chapter, there is evidence that the rate of both our boredom and our anxiety accumulation while engaged in information foraging is actually increasing in recent times, seemingly in direct response to modern technology: we are getting bored with what we are doing and anxious to move on more quickly than ever before. This contributes to a pervasive pattern of shallower resource intake curves, and thus more frequent media multitasking behavior. Under these conditions, switching would take place even if there were remaining tasty information treats to be consumed at the original source. In other words, internal factors of boredom and anxiety influence the *perceived* benefits of being in a patch, even if only subconsciously, to offset the value of consuming important information in a sustained manner.

But it doesn't end there. This shift on the right side of the model interacts with another shift that has occurred on left side of the model, *also* because of an influence of modern technology. The left-side shift is a decrease in the expected transit time to reach a new source as a result of a dramatic increase in accessibility to new information patches, particularly because of one of our game changers—the smartphone—which offers an infinite supply of

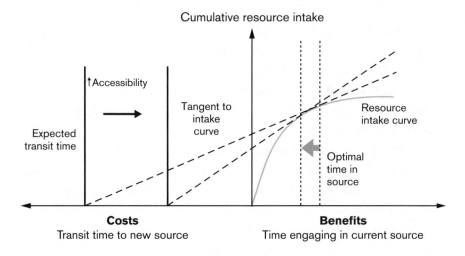

Figure 9.3

Increased accessibility to information have decreased the perceived transit time to a new source and resulted in more rapid task switching.

beckoning patches that sit in our pockets and are available at the tap of an icon. This results in even more rapid switching between information sources, despite the negative consequences associated with this behavior.

In summary, we propose that these influences of modern media and high-tech—increased rate of accumulation of anxiety and boredom, and increased accessibility of information—have driven the behaviors in humans described throughout Part II of this book. This in turn has exacerbated the collision between our lofty goals and our cognitive control limitations, as described in Part I, to result in the many negative consequences in our lives that we have been discussing. Let's examine in more detail these influences on both sides of the MVT model.

BOREDOM

It seems clear, especially when it comes to young adults, that boredom sets in very rapidly while information foraging. Stanford University professor Leo Yeykelis and his colleagues outfitted the computers of twelve students with a device that took screenshots every few seconds to determine how much task switching a typical student does during a single ten-hour day while in his or her home environment. Yeykelis found that, on average, students spent only sixty-five seconds on one screen before switching to another, but even more surprising, half of the switches occurred within nineteen seconds.[4] That means that these Stanford University students were switching screens roughly five times a minute.

In addition to monitoring screen attention, Yeykelis and his colleagues fitted each student with a special wrist sensor that measured arousal level continuously during that time period using galvanic skin response (GSR), which shows arousal level but does not indicate the source of that arousal. GSR has been used to measure anxiety and stress and also to measure excitement, so any conclusions about its meaning must be taken with caution. With that caveat in mind, the sensor data added some details to the picture and suggested why the students might switch screens so often. Across all switches, Yeykelis and his colleagues discovered that the arousal level started rising an average of twelve seconds prior to a switch but, more importantly, early arousal was most prominent when switching from "work-related" screens such as word processing or Internet information searches to "entertainment-related" screens including watching a video, gaming, and, of course, Facebook. In fact, while looking at work-related screens, arousal was quite low and the anticipatory increase was quite pronounced as the student prepared to leave the boring schoolwork and find something more stimulating such as an entertainment-related screen. In those cases, the arousal increase started nearly thirty seconds prior to the switch, showing that the students were in what the researchers called a "hunting" state that led them to something decidedly more fun (and less boring) than schoolwork. Face it: work and schoolwork are not always entertaining. But video games, social media, online videos, and electronic missives are much more interesting and rewarding, and we are often driven to a more entertaining source without our conscious realization of what is happening.

Boredom, as a field of psychological study, has not received much attention, perhaps because it is difficult to define accurately what it means to be bored. For example, John Eastwood and his colleagues at York University suggest in their work that boredom is "the aversive experience of wanting, but being unable, to engage in satisfying activity."[5] Eastwood and his colleagues go further in clarifying boredom as an aversive state that occurs when we

- are not able to successfully engage attention with internal (e.g., thoughts or feelings) or external (e.g., environmental stimuli) information required for participating in satisfying activity,
- are focused on the fact that we are not able to engage attention and participate in satisfying activity, and
- attribute the cause of our aversive state to the environment.

Eastwood further clarifies that a bored person is not just one who has nothing to do; rather, he or she wants to be stimulated and is unable to be. He calls it having an "unengaged mind."[6]

Others have viewed boredom similarly but often from a different psychological perspective. One researcher defined it as "a state of relatively low arousal and dissatisfaction which is attributed to an inadequately satisfying environment,"[7] while another suggested that it is "a restless, irritable feeling that the subject's current activity or situation holds no appeal, and that there is a need to get on with something more interesting."[8] Finally, a third definition, gleaned from the work of Erich Fromm, asserts that "boredom is anxiety about the absence of meaning in a person's activities or circumstances."[9] As these range from cognitive to affective motivations and from internal to external causes, it is not surprising that it has been difficult to pin down a perfect definition of boredom. For our purposes, we will treat boredom as a combination of both internal and external motivations and causes and further delve into the role technology plays in boredom.

First, however, it is important to note the prevalence of boredom and how it might be related to interruptions by technology. A recent Nielsen study monitored cell phone use of 3,743 American adults and asked about their motivations for using smartphone apps.[10] The three top reasons: while alone (70 percent), when bored or killing time (68 percent), or while waiting for something or someone (61 percent). All of these situations represent the possibility that the motivation for using an app is having free time and not being stimulated or not having something to do. A similar study from the UK reported that 52 percent of 1,350 adult smartphone users and 62 percent of those between eighteen and thirty reported that they prefer to use their device rather than just sit and think.[11] Finally, a cross-cultural study of thirteen countries including first- and third-world nations found that across all nations 34 percent of the 13,000 respondents said they visited social networks, played games, texted, or instant messaged when bored.[12] Although this may not seem staggeringly high, the study replicated another that was done just three years earlier and found that the percentage of people who preferred to grab their smartphone rather than be bored had doubled. Which countries showed the most smartphone activity when bored? North American countries and Asian countries topped the list. A recent study from Dr. Rosen's lab asked middle school, high school, and university students why they preferred to switch from one task to another before the first was completed.[13] Getting a text message was the most popular response (68 percent), followed by boredom (63 percent), both of which represent the two causes of interruption: external and internal.

One possibility for why the rate of boredom accumulation has increased in recent times is the influence of pervasive short timescale reward cycles in modern media. From decades of research on learning and behavior, we know that the shorter the time between reinforcements (rewards), the stronger the drive to complete that behavior and gain the reward. Take video games, for example: they are incredibly immersive and engaging, and for many games the rewards occur at extremely rapid timescales, sometimes as many as one reward per second. In a very popular mobile game, *Temple Run*, players rapidly receive coins each moment of game play (several a second) as they blast along a path being chased by a beast. It is worth considering that frequent exposure to such rapid rewards in video games may alter the boredom profile that is experienced when we engage in less stimulating information such as reading websites, which have considerably longer time-scale reward structures. There is also no reason to believe that this influence should be limited to video game play; think about the timescale of reward (i.e., information bursts) when texting. Rapid media multitasking itself, with such a high novelty load that induces frequent reward feedback, may also influence the boredom curve. In other words, this may all be cyclical: boredom drives frequent switching to new tasks → rapidly induced rewards → increased rate of boredom in nonstimulating information sources → rapid flattening of resource intake curve → quicker switch times → and so on.

Another explanation for the relationship between boredom and switching harkens back to the original learning theory work of B. F. Skinner. Skinner described the concept of "intermittent reinforcement" and showed that when someone is reinforced only some of the time, and particularly when that occurs on a variable (unpredictable) schedule, the behavior itself becomes resistant to extinction. Think about the many forms of information that might provide intermittent reinforcement inside our smartphone. Email and social media are two clear examples. When you check your email, you will probably ignore or delete many of them, but there is always at least one gem, an email that brings some pleasure. The same can be said for checking any of your social media accounts. Again, much of it may be skimmed and deemed uninteresting but there is always a post or two that are interesting and thus get read (and even "liked"), which enhances positive feelings. Clearly the more ways we have to connect to information and communication on our smartphone, the more likely it is that switching from one information patch to the stimulation from the information contained in our smartphone simply reinforces the cycle. And so, the next time we are bored, our past experiences, having gained reinforcement from our smartphone, will drive us to self-interrupt even an important source of information, and even our own quiet time.

Two recent studies by Andrew Lepp and his colleagues from Kent State University found that college students who were "heavy smartphone users" (ten hours per day) commonly mentioned boredom as motivation for cell phone use and were more susceptible to leisure boredom than "low smartphone users" (three hours per day).[14] In another study, psychologists at the University of Waterloo found that university students who were more susceptible to boredom tended to overestimate the passage of time (and were more variable in their estimates).[15] This suggests that one possible cause of boredom is that a minute can seem like an hour if you are prone to being bored. Teresa Belton, an educator from the University of East Anglia, further displayed concern for children, our most avid of the high-tech users, when she opined, "Whenever children are bored, they're likely to turn on one of those electronic things and be bombarded with stimuli from the external world rather than having to rely on internal resources or devise their own activities."[16] In fact, according to John Eastwood, it's not only students who use their smartphones to stave off boredom—we all do:

> In today's electronic world, it's rare to be stuck with absolutely nothing to do. Most of us are bombarded by near-constant stimuli such as tweets, texts and a seemingly limitless supply of cat videos right at our fingertips. But all those diversions don't seem to have alleviated society's collective boredom. The reverse may be true. These might distract you in the short run, but I think it makes you more susceptible to boredom in the long run, and less able to find ways to engage yourself.[17]

We see that the impact of boredom is not just to make us switch between information patches; we also seem to have lost the ability to simply do nothing and endure boredom. This leaves little time for reflection, deep thinking, or even just simply sitting back and letting our random thoughts drive us places we might not have gone while immersed in directed thinking. According to researchers at the Oxford, England Social Issues Research Center:

> Informational overload from all quarters means that there can often be very little time for personal thought, reflection, or even just "zoning out." With a mobile (phone) that is constantly switched on and a plethora of entertainments available to distract the naked eye, it is understandable that some people find it difficult to actually get bored in that particular fidgety, introspective kind of way.[18]

ANXIETY

The other major internal factor that influences the right side of the MVT model to more rapidly flatten the resource intake curve and leads to premature information patch switching is anxiety. Amazingly, anxiety disorders have increased twentyfold in the past thirty years. One in five people suffered from an anxiety disorder in the last year, and 28.8 percent will suffer from one at some point in their life.[19] More seriously, half of all young people, between eighteen and thirty-two, suffered from an anxiety disorder during the past year. How much of this can be attributed to technology use, and how might that anxiety influence our propensity of interrupting our attentional focus?

A recent study from Dr. Rosen's lab found that nearly half of the younger iGeneration and Net Generation reported feeling moderately to highly anxious if they could not check their text messages at least every fifteen minutes, and substantial numbers also reported feeling anxious if they could not check in with other technologies—including social media, cell phone calls, and email—as often as they would like.[20] This constant anxiety has been dubbed FOMO—fear of missing out, or the pervasive apprehension that others might be having valuable experiences from which one is absent.[21]

Although FOMO may sound rather trite, it is a real phenomenon and reflects deep feelings of anxiety. For example, one study found that in general younger adults showed more FOMO, and those who did spent more time engaged with social media and were more likely to check social media sites immediately upon awakening, right before going to sleep, and during meals.[22] Another national study by the Harris polling organization found that 40 percent of adults and 54 percent of young adults (18 to 34) would rather undertake unpleasant or even possibly painful activities—ranging from waiting in line for the DMV to sitting in traffic for four hours to getting a root canal and even spending a night in jail—before giving up their social media accounts.[23] According to Jeff Tingsley, CEO of MyLife, a company dedicated to protecting online privacy, "consumers are bombarded with so much information online ... that our anxiety around 'missing out' has shifted to digital lives [reliance on social networks]. ... Many people would rather run a marathon or spend a night in jail than give up their Facebook or Twitter account."[24]

The power of social media to increase our anxiety has been shown in numerous studies. In a study by Dr. Rosen's lab described in chapter 7, we examined which technology and media sources predicted certain psychiatric disorders.[25] In this study of more than 1,000 adults, we found that clinical symptoms of OCD—obsessive compulsive disorder—were predicted by anxiety about not checking in with smartphones, by a constant need to switch from one task to another even before the first task was completed, as well as overall use of social media. Interestingly, in that same study, those who used social media more experienced greater symptoms of OCD regardless of their level of anxiety. Other studies have shown similar results with social media driving constant checking-in behavior. As a population, it would seem that we feel an increasing need to stay connected, and this "obsession" compels us to check in with our technology often—to the detriment of our ability to stay focused on what we are doing in the moment—and when we cannot check in as often as we like (or are driven to), we feel anxious.

It is not just anxiety surrounding social media, however, that is driving us to self-interrupt. In a larger sense, our smartphone is a major contributor to our anxiety. In one survey of 3,800 adults, Cisco Systems reported that nine in ten adults under the age of thirty fear not having their mobile phone.[26] This was corroborated by a T-Mobile study showing that nearly half would miss their phone if it was out of their possession for an hour or less, and 55 percent of women would rather leave home without makeup than leave home without their phone.[27] A UK-based study found similar results among 1,000 British adults, where 66 percent (and 77 percent of young adults) feared losing or being without their phone, which was an increase of 13 percent in a similar study performed just four years prior.[28]

In a study discussed in detail in chapter 8, we highlighted the anxiety raised by taking away someone's phone.[29] The impetus for the study was the hypothesis that taking away someone's phone would make him or her highly anxious, even more so than having a phone nearby (albeit turned off). The results showed that this hypothesis was wrong. On average, anxiety increased across the hour for *both* groups. But that's not the whole story. Given that there was a range of technology use, the users were segmented into light users, moderate users, and heavy users by simply dividing them into thirds based on their daily use of several forms of technology including smartphones, Internet, video games, and so on. Anxiety levels did not differ for any of these technology user groups except for one: smartphone users. Light smartphone users—that is, those who check their phone here and there during the day and were not uncomfortable if they couldn't check in—showed absolutely no change in anxiety across the hour. The moderate users, however, showed an initial increase in anxiety that leveled off at a moderate level of anxiety. The heavy smartphone users, however, demonstrated an entirely different picture. First, even at the initial testing after only ten minutes without being able to use their phones—regardless of whether the phone was under the desk or taken away—they already showed heightened anxiety, which increased by leaps and bounds over the rest of the testing time. Who were the heavy users? Mostly members of the iGeneration and the Net Generation, but there were also older students who were telling us that they were anxious. Out of sight was clearly not out of mind. And as we have seen, the typical smartphone user has rapidly become a frequent user, keeping the phone close at hand all day and all night.

Dr. Rosen's study of cell phone separation evaluated anxiety when the students were doing nothing, but another study of iPhone users looked at anxiety when they were engaged in a task (an information patch) and also extended the assessment to both paper-and-pencil tests and psychophysiological measures of anxiety. Lead author Russell Clayton, of the Missouri School of Journalism, brought forty iPhone users into the lab one at a time and had them perform word search puzzles under two conditions: with their iPhone close at hand but silenced and later on with their iPhone moved about four feet away in plain sight and sound.[30] Subjects were connected to devices that monitored their heart rate and blood pressure throughout the tasks, both when their phone was nearby and when it was moved away. At a specific time after the iPhone was moved, the experimenters surreptitiously rang the iPhone but did not allow the subject to answer the phone call. Both heart rate and blood pressure increased, as did self-reported anxiety following the missed call. Other researchers, both in the United States and other countries, have found similar effects in a variety of studies using a medley of assessment tools: those who use their cell phone more during a typical day show more anxiety.[31] One fascinating New York Times article explored the issue of "typing awareness indicators" defined as the ellipses that indicate that a fellow iPhone user is writing a text message or the "so and so is typing" indicator on many instant messaging and chat sites.[32] The article examined the anxiety we face while awaiting a message and related our tendency to ignore all else around us and fixate on the indicator, waiting for an incoming message, to "watching paint dry."

So, why have anxiety levels changed for so many of us when we engage in information foraging and also when denied access to our favorite information patches? We believe that the main aspect of modern technology that has contributed to this shift is that all three game changers increased not merely our access to information, but notably to a particular type of information: communication. The ability to connect with other people, anywhere, all the time, has changed dramatically with the advent of email, mobile phones, texting, social networks, and myriad ways for one person to connect to another electronically. We hypothesize that this has put pressure on our expectations of interconnectivity and that has resulted in the emergence of heightened levels of anxiety accompanied with FOMO, nomophobia, and phantom pocket vibrations.

With so much anxiety being generated, it is hardly surprising that many of us are feeling compelled, some almost to the point of an obsessive-compulsive disorder, to engage with our technology. As a population, it would seem that we feel an increasing need to stay connected, and this "obsession" compels us to check in with our technology often—to the detriment of our ability to stay focused on what we are doing in the moment. And when we cannot check in as often as we like (or are driven to), we feel anxious. Note that there may be other causes of heightened anxiety to move on to other patches aside from social ones—for example, motivation for high performance and productivity in general.

ACCESSIBILITY

We have discussed the two main forces that flatten the resource intake curve on the right side of the MVT model to drive the intercept toward an earlier optimal time in source, thus resulting in us shifting our attention more often. But it is important to appreciate that there is a major factor on the left side of the model that decreases the expected transit time to a new source, resulting in an even further shift of the time in source to an earlier time point. That factor is *accessibility*. The more readily available a new patch of information is (or even seems to be), the earlier the time someone will disengage from his or her current source. It is essentially the same for animals foraging for food; if another tree full of nuts is sitting right there, a squirrel is going to take the leap earlier to this new patch than if it has to travel far to get to another nut-filled tree.

Here again, we find a powerful influence of media and technology. Information has never been more accessible. Our computers provide pop-ups, email alerts, number of unread texts and tweets, pending chat messages across multiple platforms, unattended social media notifications, calendar alerts, pending software updates, and, on some computers, even constantly bouncing alerting icons. As soon as we unlock our device from its resting state we are presented with numerous apps that beckon for our attention and remind us how accessible that next patch is. We now park ourselves in front of those multiple screens including our small phone screen, our intermediate-size tablet screen, our somewhat larger computer screen, and our high-definition large television screen. And each one carries the promise (and peril) of multiple windows and tabs all designed to lure attention away from our current task to something that promises information that may be more interesting. Links to ever more content fill every screen. The concept of surfing the Internet is quite a good one; we catch wave after wave of information patches with the click of a mouse. And technology offers an intense sensory experience that includes stimulation of all senses by a variety of devices. Given that we all possess—and have close at hand—many devices that signal our attention, we may leave our current patch not because we have finished our resource intake at that location but simply because we have been alerted through stimulation of our sensory system by strong bottom-up influences that something more interesting or intriguing is available.

Most of us carry an accessibility portal in our pocket or purse at all times, including, as we have shown earlier, close at hand while we sleep. Smartphones operate everywhere and at lightning speed. Wi-Fi, at ever-increasing speeds, is available in nearly every location imaginable. We watch television with our smartphone or tablet close at hand. We are rarely without at least one device 24/7/365. They join us in bed and on vacation, places that did not feature these types of interruptive devices just a short time ago. Everything features links that we simply tap or click and are somewhat magically taken to a new location that might offer something more "entertaining."

Of note, the accessibility is not just our ease in reaching out to our technology to forage information. It is also the very powerful way in which *technology can now reach out to us*. This truly changes everything. It would be as if a neighboring tree could go ahead and throw a nut at a squirrel any time it was interested in being fed upon. These interruptions dramatically alter the expected transit time to a new source because it constantly reminds us just how accessible it is. As Cory Doctorow so aptly observed when writing an article for *Locus Magazine* on why he finds it has become so difficult to write these days, "The biggest impediment to concentration is your computer's ecosystem of interruption technologies: IM, email alerts, RSS alerts, Skype rings, etc. Anything that requires you to wait for a response, even subconsciously, occupies your attention. Anything that leaps up on your screen to announce something new, occupies your attention."[33] Doctorow referred to our behavior amid the constant temptations to an "endless click trance" that can turn just a few minutes of writing into hours of distractions.

In summary, technology-powered influences on the right side of the model—boredom and anxiety—interact with a technology-powered influence on the left side—accessibility—to result in a shifted intercept between them, which is expressed as a much earlier "optimal time in source." There seems no denying that these influences on both sides of the model are associated with an earlier switch time, but is it really an optimal one? In terms of the big picture, it is now clear that for the most part the answer is no. There are just so many negative consequences to this behavior, especially when it is frequent. But the model was not formulated to account for these powerful internal factors; it was developed to specifically understand optimal foraging for resources as dictated by external influences in animals that are instinctively optimizing their survival. Even if we evaluate this in terms of the goal of optimal foraging of information, it still seems that our behavior is not optimal—that these technology-influenced factors work together to generate behaviors that are not even ideal for information foraging. Too much important information is being "left on the table."

METACOGNITION

On top of the direct impact of changes in anxiety, boredom, and accessibility induced by modern technology on the Distracted Mind, there also seems to be an important role played by poor introspection into our own minds and its vulnerabilities and how this may affect our performance. This lack of metacognition—awareness and understanding of one's own thought processes—impacts the MVT model in two ways. On the right side: not understanding the benefits of remaining in an information patch *and* not appreciating our internal states of anxiety and boredom. On the left side: not accurately evaluating the consequences of moving to a new patch, that is, a lack of understanding of the performance costs of multitasking and task switching. Many people believe that we are more productive if we spend "just a few moments" dealing with that incoming message or searching for that tidbit of information, rather than sustaining our attention on a task, and resisting distractions and the allure of interruptions. As described in previous chapters, this belief is misguided and causes countless problems both in terms of productivity and in preserving our physical and mental health. The truth is that we are mostly oblivious to the toll that constant task switching generates. We convince ourselves that we can handle it because we mistakenly believe that we possess a brain that is built for multitasking; or, because we do it all the time, we feel that we must have become really good at it.

As evidence, it has been shown that people who believe that they are good at multitasking actually tend to be those who do the worst on laboratory tests of multitasking, leading the study authors to conclude that "participants' perceptions of their multi-tasking ability were poorly grounded in reality."[34] Moreover, those individuals who actually media multitask the most, and self-report the most cell phone usage while driving, are actually the least equipped to successfully pull this off, as documented by poorer performance on cognitive control assessments of task switching, multitasking, and distraction resistance.[35] In addition, studies have shown that when asked to estimate the costs of doing two complex tasks at the same time, most people will understand that doing them together will take more time, but there is a zero correlation between the predicted and actual costs of multitasking, showing that we really don't have a handle on exactly how detrimental to performance it is to attempt to do two tasks at the same time.[36] Research from driving studies, for example, shows that when people are given control of when to perform a secondary task while driving in a simulator, they do not take into consideration the ease or difficulty of the current driving conditions, but instead act as though the incoming text message, email message, or phone call takes priority regardless of whether it is advisable.[37]

Many of our actions regarding task switching are subconsciously driven by the factors of the optimal foraging model, and so are not actual decisions. However, sometimes we do make decisions. You may have made a calculated evaluation of what you might gain by, say, texting while driving, and decided that the potential risks are outweighed by the anticipated outcomes. You may have even heard that we cannot text and drive very well, and yet you do so regardless of the knowledge that if caught you'll have to pay a large fine, and even more critical, that it might lead to an accident and possible injury or death to you or others.[38] In psychological and educational terms, we are displaying poor metacognition or a lack of understanding of how our mind works and why our behavior might not be the ideal given our innately limited human capacity for doing more than one attentionally demanding task at a time. And we see this lack of insight manifested in multiple ways. According to researchers at Microsoft and the University of Illinois:

> Even though users feel that they are in control of when they switch tasks due to an alert, they appear to be largely unaware of the amount of time they end up spending on the alerting application, on other tasks they invoke as a result of responding to the alert, and on browsing through other peripheral applications before resuming the suspended task. Even when users respond immediately with the intention of resuming the suspended current task as soon as possible, they often end up taking significantly more time to return than the time to respond.[39]

This lack of metacognition was also evident in a recent study out of Washington University.[40] Researchers asked students to perform two difficult computer tasks, one at a time, which included keeping their cursor on a target that was moving around erratically on the screen and the "n-back" task, which involves watching a sequence of letters on the screen and monitoring whether a letter is repeated next (zero-back), repeated with another letter in between (one-back), repeated with two letters in between, and so on. The participants did indeed estimate that doing both the n-back task and following the target at the same time would be more difficult than doing them one at a time, evidencing that metacognitively they understood that this burden on their cognitive load would negatively affect their performance. However, there was no consistency in their rating of the difficulty and their actual performance in doing both tasks at the same time. Yes, they knew that doing more than one task at the same time would be more difficult, but they had no conception of how difficult it would be. In reconciling the consistency of their results with similar studies of driving behavior, the study authors suggest that "it could be that drivers rely on *relative* judgments to guide multitasking behavior (e.g., I'm better than the average driver at handling distraction, so I'll answer this phone call)." The authors further suggest that perhaps drivers do possess metacognitive judgments about the difficulty of texting and driving, but that they assess the specific situation as warranting the distraction.

In Part II, we discussed how we are overwhelmed by interference from both outside and from within, and how this one-two punch has driven us away from sustained allocation of our attention to a whimsical whipping of our attention away from what we are should be doing toward something that promises to be better, or at least more enjoyable. Our ability to maintain our focus in one information patch—whether it be a work project, a homework assignment, or something as simple as watching a television program—has been seriously jeopardized, and we believe that modern information technology is a major culprit. We have become such massive consumers of high-tech, spending the vast majority of our waking hours using one or more devices, and we have not really had much practice sustaining our attention on just one activity and ignoring the lure of the others. We do not think that this is a calamity from which we cannot rescue ourselves. In fact, we believe that technology—the same technology that has captured our attention—offers a variety of ways to help is regain our focused attention and retrain our brains to withstand the beckoning fingers that drag us away from our goals. In Part III, we explore solutions and strategies that you may adopt to help maintain focus and thus aid your Distracted Mind.

Part III

TAKING CONTROL

THERE ARE TWO approaches by which we can diminish the negative impact of interference on our lives: changing our brains and changing our behavior. In this final part of the book, we will present our latest perspectives on both approaches and provide practical advice for taking control of our Distracted Mind. Note that these approaches are not mutually exclusive; they are complementary, and you will likely achieve the most beneficial outcomes if you pursue them concurrently.

In terms of changing our brains, laboratories and companies around the world are now engaged in large-scale development and research efforts directed at understanding how we can enhance our brain's functioning to improve cognitive control and thus reduce the negative impact of goal interference. In chapter 10, we review the evidence for the many approaches of cognitive control enhancement: traditional education, meditation, cognitive training, video games, exposure to nature, drugs, physical exercise, neurofeedback, and brain stimulation. Interestingly, many of them use modern technology to harness neuroplasticity and induce brain changes. We are at the threshold of fascinating times, as the technology that has aggravated the Distracted Mind is now being formulated to offer remediation. Research in this domain is still in its infancy, but we will take you on a tour of many interventions that may change our brains, so that when we have no choice but to engage in a high-interference environment, we are as optimized as possible to diminish the detrimental effects of distractions and interruptions.

It may seem difficult to change our habits, but at many points through-out history we have decided to modify how we interact with our environ-ment after we've realized the harmful effects of a particular behavior. For example, prior to the overwhelming evidence of the dangers of cigarette use, doctors would actually recommend specific brands in advertisements. With increasing knowledge of the detrimental effects of this sort of behavior, and others such as the dangers of sun exposure, came the opportunity for us to make more informed decisions. In chapter 11, we provide practical advice and strategies on how you can modify your behavior as you interact with both high-tech and low-tech to minimize the impact of goal interference. This advice, based on the MVT model, will serve as a framework for under-standing how you can positively bias the impact of technology on your life. Strategic approaches to gaining control will focus on mitigating the negative influence of these four issues that arise from the MVT model: poor metacog-nition, increased accessibility, boredom, and anxiety.

10 BOOSTING CONTROL

And the faculty of voluntarily bringing back a wandering attention, over and over again, is the very root of judgment, character, and will. No one is *compos sui* if he have it not. An education which should improve this faculty would be *the* education *par excellence*.
—WILLIAM JAMES[1]

AS DESCRIBED IN PART I, our highly evolved goal-setting abilities collide headfirst with fundamental limitations in our cognitive control to generate goal interference that manifests itself as the Distracted Mind. If we want to improve our quality of life, one option is to strengthen our cognitive control by minimizing its limitations, thus reducing goal interference. Is this possible? Can we actually enhance our brain's functioning to improve cognitive control? Many scientists believe that we can. The fact is that our brains are always changing. One of the most fundamental features of our brain is its very ability to change, a phenomenon known as *neuroplasticity*. It is now well understood that the brain modifies itself at every level, from structure to chemistry to physiology, in response to interactions with the environment. This is the very basis of all learning, and we now appreciate that plasticity does not end after critical stages of development, as was once believed, but persists throughout our entire lives. The challenge we face is to figure out how to maximally harness our brain's plasticity to accomplish this goal and then carry out careful research studies to validate that change is taking place and that it has the desired consequences.

We now appreciate that brain changes induced by neuroplasticity can endure and harbor benefits many years later via another brain phenomenon known as *cognitive reserve*.[2] Building a stronger brain may even delay the negative functional consequences of degenerative neurological diseases, such as Alzheimer's disease.[3] And yet, just because our brains exhibit neuroplasticity and cognitive reserve, this does not mean that it is trivial to induce meaningful and sustainable changes. Our brains also have mechanisms for maintaining *homeostasis*, a state of stability that is critical for our survival; can you imagine how deleterious it would be if our brains changed dramatically in response to even casual influences?

Let us consider the entire landscape of approaches that have potential to enhance our core cognitive control abilities. First, there are approaches that, although distinct in their implementation, are similar in that they expose individuals to specifically tailored environments, interactions, and experiences that stimulate brain plasticity. These include both traditional and modern approaches of education, meditation, nature exposure, cognitive exercises, video game training, and physical exercises. Second, and by far the most widely engaged approach to impact cognition for individuals with deficits, is the ingestion of designer molecules, also known as drug treatment. Third, there are interventions—neurofeedback and brain stimulation—that may seem more natural in the world of science fiction but are actually active research areas.

It is helpful to view the evidence for an approach being an effective tool in enhancing cognitive control as falling into one of three levels. The highest level is generated by positive research findings from randomized, placebo-controlled, double-blinded studies, known as "randomized controlled trials" or RCTs. This type of study design is essential to minimize biases that result in false and/or narrow conclusions. This is the standard required to establish advice as *prescriptive*. Ideally, RCTs involve large numbers of participants, move beyond laboratory measures to assess real-world impact, evaluate the magnitude of positive and negative effects, and are replicated by several independent studies. The next level down the ladder is positive evidence that is considered a *signal*. This is when careful laboratory research studies establish feasibility, mechanism, and proof-of-principle that something statistically significant is going on, but there is still the need for large-scale RCTs to advance the evidence to the prescriptive level. Research studies that offer us a signal are critical for generating interest as well as the resources to move the approach to the next level. The lowest level of evidence is from research studies that are positive, but really serve as the basis to generate a *reasonable hypothesis*. This is when some evidence has accumulated, perhaps from anecdotal reports and a several research studies, plus sound logic based on the scientific literature, which suggests that an approach is worthy of deeper investigation.

We will share with you our impressions of the level of evidence currently achieved for each approach to enhance our cognitive control: *prescriptive, signal*, or *reasonable hypothesis*. You can use this to guide you in making informed, evidence-based decisions about which are most interesting to you. Of course, there are many factors to consider when making a decision about engaging in any intervention, which extend beyond the level of evidence. For example, there is the state of your Distracted Mind at the time, which establishes how seriously you need to use an intervention in the first place. And each approach has numerous other factors to consider, such as cost, time, magnitude of gain, negative side effects, and positive side effects such as general health improvement, stress relief, and social returns. Everyone should reach his or her own personal decision. For some, an approach that is a reasonable hypothesis, but also fun and with limited side effects, may be enough to justify its use; others may demand prescriptive evidence before engaging in any intervention that will divert time away from something else.

Let's review the evidence for each approach, focusing on their effectiveness as cognitive control enhancers with the ability to minimize our limitations and ameliorate our Distracted Minds.

TRADITIONAL EDUCATION

The underlying goal of all educational systems is "to transmit to a next generation those skills, facts, and standards of moral and social conduct that adults consider to be necessary for the next generation's material and social success."[4] The most widely implemented approach is our current system of didactic classroom instruction delivered by a teacher lecturing to a group of students. Although this long-established, globally adopted, traditional education system varies in its details by geography and historic time period, a common feature is the emphasis on rote memorization via formalized and structured lessons followed by assessments of attained knowledge using formalized testing. One issue we face is that academic performance itself has often become an end in itself rather than a means to something greater. A parent might ask his child how she is doing in math class. But often what is really being asked about is what grade she is getting in the class. Parents' thoughts are frequently occupied by how a grade will facilitate their children getting into a university, and not on whether they are sufficiently learning the intricacies of manipulating numerical information that will benefit them throughout their lives. Nor is the emphasis on the development of their cognitive control abilities.

There seems to be a tension between this traditional model that has largely focused on the delivery of *information content* and the goal of developing the core *information-processing abilities* of the brain. We believe that the objectives of an education system should not be directed solely at the transfer of content to young minds. Of course, this is important; there is so much that needs to be learned. But it is also critical that developing minds build strong cognitive control abilities that allow them to engage flexibly in dynamic and challenging environments. Even new "alternative" educational systems that aim to foster real-world outcomes by employing hands-on, project-based activities and student-led discovery encourage learning strategies related to specific skills but not necessarily the development of cognitive control abilities. While learning factual knowledge and acquiring practical skills are both important, the development of basic cognitive control abilities is essential. There is convincing evidence that superior cognitive control is associated with successful academic performance, but little is known about the reverse; that is, does traditional education actually build the fundamental information-processing abilities of our brains that underlie cognitive control?[5] Here, we address the question of whether traditional education is truly an effective form of cognitive enhancement that has the power to minimize our control limitations.

More simply put, does our current education system help the young Distracted Mind?

It is abundantly clear that the world has benefited in many ways from the global spread of traditional education. In addition to the undeniable value of increased basic literacy and numeracy skills, there is evidence that education is a major predictor of positive health outcomes.[6] The adoption of traditional education has also been closely associated with a progressive rise in the mean IQ across North America, Western Europe, and Japan over the twentieth century, a phenomenon known as the Flynn effect.[7] Research that has compared schooled and unschooled adults from farming communities suggests that the benefits of schooling may indeed reach beyond the accumulation of factual knowledge to the development of specific skills to the attainment of better reasoning abilities, cognitive abstraction, and problem solving.[8] But, unfortunately, to the best of our knowledge there is no convincing evidence that traditional education (or for that matter "alternative" education) actually enhances our cognitive control abilities.

The evidence level that traditional education directly engenders enhancement of cognitive control seems to be just at the level of a *reasonable hypothesis*. Sure, it is logical that any structured education system (even one directed only at memorization) compared to *no* education at all will result in better cognitive control, but that seems to be a low bar. There are several actions we can take to improve our educational approach in this regard. First, we can start rigorously assessing cognitive control abilities of children, not just in the context of suspected learning disabilities, but all children. This will result in a better understanding of the strengths and weakness of each child in terms of his or her attention, working memory, and goal management, and a better understanding of how these abilities map on to their standard academic performance metrics.

The next step is to perform careful studies of novel educational approaches that are directed at boosting control abilities. This will be necessary to reach the highest level of confidence that our approach to education is best serving our children's Distracted Minds. As an example, an exciting effort to assess a new educational curriculum explicitly directed at improving children's cognitive control abilities was undertaken for the *Tools of the*

Mind (Tools) program.[9] Psychologists Elena Bodrova and Deborah Leong developed *Tools* based on theories and insights into how a system of activities can be designed to boost cognitive control.[10] It involves support and training of specific control abilities using external aids, interactions, and play that are woven into traditional teaching classroom activities. More details can be found online.[11] In the early 2000s, a research study was performed in a low-income school district in the northeast United States that followed more than one hundred preschoolers for a one- to two-year period to compare *Tools* with a curriculum that used a more traditional educational approach. The study revealed that *Tools* resulted in enhanced performance on tests of cognitive control in all three domains, which exceeded those of children exposed to the traditional curriculum. Notably, the children in the *Tools* classroom exhibited diminished distraction effects on testing, a sign that they may have indeed developed less Distracted Minds. We would classify this research finding as starting us along the path of a *signal* that new educational approaches can directly target cognitive control enhancement in a fun, interactive way in standard public school classrooms. We encourage continued innovative efforts in this direction.

We also need to think increasingly about education as a lifelong process; we have the potential to enhance our cognitive control at any age. Educational programs across the lifespan directed at boosting and maintaining cognitive control should be the rule, not the exception.

MEDITATION

Meditation is another approach that offers an organized program of interactions and experiences to enhance our minds. Meditation is not a single entity; rather, it is an umbrella term that encompasses dozens of distinct approaches. Many modern meditation approaches were born out of Buddhist traditions that have been practiced for thousands of years. The specific goals vary across the different techniques, from relaxation and stress management, to improved attention abilities, to the achievement of a heightened sense of well-being, compassion, wisdom, and altruism. We now have a growing body of evidence that meditation practices are associated with a wide range of health benefits.[12]

But what about the potential of meditation to improve cognitive control abilities? It seems logical that this may indeed be a positive benefit since many meditation practices at their core are essentially attention training. For example, a commonly adopted meditation approach is the practice of mindfulness techniques to develop skills of focusing on the present moment. Surprisingly, we do not have as much evidence about this benefit of meditation as we might hope, although this is changing rapidly.[13]

Studies to assess the potential cognitive control benefits on naïve participants engaging in meditation began to emerge in the mid-2000s. This was an advance from previous studies that analyzed the brains and minds of meditation experts, which, while fascinating, had major interpretive challenges because of the many unique factors that distinguish these individuals beyond their dedication to meditation practices. Research efforts have focused largely on the benefits of two types of meditation programs: meditation retreats that may last months, in which intensive mindfulness skills are practiced for as many as ten hours a day, and group-based programs, such as the clinically oriented Mindfulness Based Stress Reduction course (MBSR), which usually involves a weekly class for a couple of months and assigned daily home meditation for thirty minutes to an hour a day. Most participants in these studies focus their training on two styles of meditation techniques, which are sometimes combined: "focused-attention" meditation and "open-monitoring" meditation. Focused-attention meditation, also referred to as "concentrative meditation," involves directing focused attention in a sustained manner on a single object, typically one's own breath. Awareness is also directed at detecting when attention to this focal point has wandered, and then redirecting it back to a sole focus on the breath. Open-monitoring meditation is quite the opposite, in that there is no single object of attention; rather, the practitioner openly monitors his or her feelings, thoughts, and sensations from moment to moment without reacting to them. Although quite different, both meditation practices are, at their core, attention exercises.

In one of the first studies of individuals with no prior meditation experience, Dr. Amishi Jha and colleagues recruited seventeen young adults to participate in a five-week MBSR course offered by the University of Pennsylvania School of Medicine.[14] Participants attended a weekly three-hour class that included both meditation practice and group discussion, and also engaged in thirty minutes of daily meditation that mostly involved focused-attention techniques, with an introduction to open-monitoring in the final week. Results indicated that participants exhibited improvements in selective attention compared to those in a control group who did not train over the same time period. This study was consistent with findings from previous research that showed expert meditators excelled on selective attention tasks compared to nonmeditators. Over the years more evidence has accrued that meditation techniques improve cognitive control, including sustained attention, speed of processing, and working memory capacity.[15] In addition, one recent study took a step toward documenting real-world impact by showing meditation-induced improvements in the Graduate Record Examination (GRE) reading-comprehension test.[16]

Accumulating evidence convinces us that there is a strong *signal* that meditation engenders improvements in cognitive control, and of course there are many reasons beyond improvements in this domain that encourage us to recommend engagement in mindfulness practices. However, many studies have methodological limitations that cause this approach to fall short of the stronger *prescriptive* levels of evidence. There are certainly enough positive results to encourage more rigorous meditation research and RCTs to better understand meditation's cognitive control benefits, and how we may harness these ancient practices to achieve a less Distracted Mind.

COGNITIVE EXERCISE (BRAIN GAMES)

Cognitive exercises, also sometimes referred to as cognitive training or brain training, have become a popular activity for many wishing to boost their brain's performance.[17] Based on a similar concept to what drives physical fitness training programs, these mental exercises usually involve engaging in repetitive and adaptive interactions with a demanding cognitive task. Often the tasks used for these exercises are derivatives of cognitive tasks used by psychologists to assess the same abilities that they are aimed at improving. The basic idea is that challenging a specific cognitive ability over time causes it to become stronger in response to task demands, much like exercising in a gym results in strengthening muscles by repetitively pushing them against resistance.

Cognitive exercises are an attempt to improve brain function by harnessing our brain's inherent plasticity, rather than by explicitly teaching a strategy or a skill. Most training programs attempt to accomplish this goal not just through repetitive task engagement, but also through *adaptivity*. Adaptivity means that as a participant's performance improves over time, the task is made more difficult gradually, conceptually similar to the role of a physical fitness trainer who increases the weight in an exercise in response to the trainee getting stronger. Cognitive exercises often now accomplish adaptivity using computerized software algorithms that adapt task challenge in real time based on recorded performance metrics. This is one of the major advantages of computerized training approaches versus static, non-computer-based training approaches. Another advantage is that computerized training allows participants to more carefully monitor their performance in real time over the course of training. When the feedback involves playful and fun motivating game elements, these exercises are sometimes called "brain games."

Enhancing cognitive control abilities is perhaps the most popular goal of commercial and academic cognitive exercises. This is not so surprising, as it has been increasingly recognized that successful engagement of cognitive control is necessary for all higher-order cognition. Many of these efforts have focused on improving cognitive control in older adults who have significant deficits in this domain that affect their quality of life.[18] But a growing number of studies are directed at the development and validation of cognitive exercises for children and young adults, as well as diverse patient populations.

Attention training has become a particularly active area of scientific research, with the goal being to determine if cognitive exercises can improve attention with positive benefits on real-life performance. In March 1998, recruitment began for thousands of adults over sixty-five years of age to participate in the largest randomized cognitive training study, known as the ACTIVE trial.[19] In addition to a memory and reasoning training group, this study involved a testing group that engaged in an adaptive, computerized, attention-training exercise for ten sessions over five to six weeks. The study revealed that participants exhibited improved attention as determined using the Useful Field of View (UFOV) test, which was still evident in follow-up studies ten years later.[20] Interestingly, a follow-up study also showed an approximately 50 percent lower rate of at-fault motor vehicle collisions relative to a control group.[21] Long-term follow-up of the participants revealed that those older adults who engaged in these attention exercises ten years earlier self-reported less difficulty in daily life activities than the control group. In spite of these promising results, the ACTIVE trial also revealed some significant limitations of this cognitive exercise, notably limited evidence of transfer of attention training to benefits in other cognitive domains and objective measurements of real-world performance.

Despite successes with attention training in older adults over the years, a critical aspect of attention—the resistance to a negative impact of distraction—has proven to be very challenging to remedy by cognitive exercises. In 2009 the Gazzaley Lab launched a project under the leadership of Dr. Jyoti Mishra to develop and evaluate a novel attention exercise called *Beepseeker*, which was designed to directly target the act of *ignoring* irrelevant information.[22] In *Beepseeker*, participants heard three tones and had to correctly identify if a "target" tone was present among the "distractor" tones. As their abilities improved over time the distractor tones moved closer in frequency to the target tone, so that the challenge was adaptive. We studied the influence of *Beepseeker* on cognitive control abilities not just for older adults, but also older rats, which exhibit similar challenges with ignoring distractions.

We found that *Beepseeker* training resulted in both older adults and older rats improving their ability to resist the negative impact of distraction. Neural recordings from both species showed that their brains had an enhanced ability to suppress distractors. Importantly, we also found that engaging in this attention exercise resulted in improvements in working memory. These results confirmed our hypothesis that adaptivity is a powerful tool to target a cognitive exercise to a specific process that needs to be improved, as well as suggesting that this exercise might also benefit other cognitive control abilities.

Studies have shown that there are other cognitive exercises outside of the domain of attention that can generate positive effects on cognitive control, and also that they are not just beneficial for older adults. Cognitive exercises designed to challenge working memory and goal management (both task-switching and multitasking) have shown evidence of some degree of generalizable improvements in cognitive control for individuals of all ages.[23] For example, a study that trained disadvantaged first graders on a combination of cognitive exercises, each targeting different abilities, showed signs of success.[24] In addition to showing improvements in cognitive control, these children also displayed improvements in language and math grades in school, which equalized their academic performance with children who attended school more frequently.

We should note that other studies have shown that cognitive exercises resulted in no transfer of improvements to nontrained tasks.[25] It is clear that we need to better understand the factors that lead to such discrepant findings if we are to continue to advance this approach as an aid to the Distracted Mind. It is likely that differences in the quality of the exercises themselves, as well as the time and depth of engagement in the training programs, contribute to different outcomes. We also need to refine the experimental methods used in the studies.[26]

The accumulation of data from dozens of studies over the last decade convinces us that there is a *signal* here—cognitive exercises that challenge specific cognitive control abilities can reduce their limitations. However, despite many successful studies, conflicting results in the scientific literature feed an ongoing debate over which specific exercises generate benefits and the degree to which these exercises result in a transfer of benefits to abilities beyond those trained directly. This conversation has been aggravated by unease among scientists by overinflated marketing claims made by "brain game" companies who have not generated sufficient data that demonstrate their specific cognitive exercises are effective. These concerns, coupled with the need for larger-scale RCTs with better experimental design, currently limit cognitive exercises in our minds from reaching the highest level of being a *prescriptive* approach. We remain cautiously optimistic that the signal is real, and with continued rigorous research elucidating the mechanism of action, transfer of benefits, sustainability of effects, and personalization factors that lead to effective results, we will one day have powerful new cognitive exercises to aid the Distracted Mind.

VIDEO GAMES

Closely related to cognitive exercises are video games, which are interactive media that are also often designed to be demanding, adaptive, and include lots of feedback, but differ in terms of several important design elements. Video games are built with fun and play as a main factor. They are designed with a primary goal of engendering high levels of immersion, engagement, and enjoyment for the players, often using complex reward structures and high levels of art, music, and story. They do not tend to focus on one specific cognitive skill, as exercises usually do, but rather expose players to multiple demands that challenge a broad range of abilities. Similar to what has emerged for cognitive exercises, an exciting field is being born to understand the impact of video games on cognition, and notably on cognitive control abilities.

The story of video games having a positive influence on the Distracted Mind began with an amazing tale of serendipity. It was 1999 at the University of Rochester when Shawn Green, an undergraduate student at the time in the laboratory of Dr. Daphne Bavelier, was preparing to initiate a new research study. Green had been working on modifying a cognitive control task, the Useful Field of View, when he ran into a perplexing issue. He noticed that he and some of his friends who were helping him with pilot testing were consistently performing off the charts on this task. Bavelier herself had tested well within the expected range of performance, raising an interesting question as to what it was about the cohort that Green had assembled that made them perform so well. A bit of investigation revealed that those who were performing at such a high level were all players of action video games, particularly games known as first-person shooters. This clue that there was a relationship between video game play and cognitive control abilities led to a groundbreaking research study that was published in the journal *Nature* in 2003 titled "Action Video Game Modifies Visual Selective Attention."[27]

This study demonstrated that when habitual action video game players are tested in the lab they exhibit superior attentional capacity, distributed attention, and speed of attentional processing compared to nonplayers. Green and Bavelier also showed that nonplayers who were recruited to play the first-person shooter game *Medal of Honor* for one hour a day over ten consecutive days improved on these same cognitive control abilities compared to another group who played the game *Tetris*. Their conclusion was that the nature of action video game play was critical in "forcing players to simultaneously juggle a number of varied tasks (detect new enemies, track existing enemies and avoid getting hurt, among others)," and this in turn reduced limitations of several aspects of cognitive control.[28]

Over the decade that followed this publication, dozens of studies have shown that playing action video games can enhance cognitive control. In addition to the various aspects of attentional enhancement shown in the 2003 paper, evidence surfaced of improved selective attention for objects, space and time, sustained attention, bottom-up attention, working memory, task switching, and multitasking.[29] Many of these cognitive control enhancements have even been shown to last for many months after game play was discontinued.[30] Neural recordings to understand the mechanisms that underlie these effects revealed that gamers exhibit a superior ability to detect targets, at least in part as a result of being better able to suppress distractions.[31] An fMRI study further revealed that gamers did not activate prefrontal cortex areas to the same degree as those who did not play video games in response to increasing attentional demands on a searching task.[32] This suggests that gamers allocate their cognitive control resources more efficiently.

In 2008, Dr. Gazzaley was inspired by these encouraging findings that pointed toward a causal relationship between action video game play and enhanced cognitive control. For many years, the Gazzaley Lab had been revealing cognitive control deficits in older adults, and we became increasingly motivated to help this population in some way rather than solely reporting on the problem. But we could not look directly to studies of action video game play for the answers, as they were all focused on younger adults. Dr. Gazzaley decided to perform a study similar to the one that had been reported in 2003, except it would be directed at healthy older adults and would not use an off-the-shelf action game, but rather would involve designing and building a video game from scratch, customized for the goal of boosting cognitive control.

The plan was to first create a novel video game and then carefully test it to see if it would improve cognitive control in the older adults who played it. Specifically, we wanted this video game to challenge older participants to perform two demanding tasks at exactly the same time while they were immersed in a distracting environment. This would fulfill our goal of engaging older adults in a scenario that is similar to what is so demanding in their daily lives: interactions that involve multitasking in a distracting landscape. The hypothesis was that if their brains gradually got better at managing this high-level interference challenge to their cognitive control, then other aspects of cognitive control that were not directly trained would also show benefits. This hypothesis was based on findings that cognitive control abilities have common underlying neural mechanisms, notably the prefrontal cortex networks that are engaged. The idea was that if we put pressure on this aspect of cognitive control, benefits would manifest in other aspects.

Given the success of action video games in boosting cognitive control in young adults, it was an appealing option to consider this approach for older adults. But why not just grab a great, off-the-shelf, first-person-shooter game that had already been used by other researchers? The first reason was that older adults are not very fond of these games, nor are they very good at them. Second, commercial action video games are not as adaptive in terms of scaling their challenge in response to improving player performance as Dr. Gazzaley felt they should be. We reasoned that adaptivity is the most critical design element for any effective plasticity-harnessing tool. Also, we wanted to control every single game element—timing, positioning, and nature of stimuli—so that we could record neural activity during game play and understand what was changing in the brain in response to training. And since we were going to build a video game from scratch, it might as well not be violent.

Later that same year, Dr. Gazzaley came up with the basic design of *NeuroRacer*, a 3D video game in which a player navigates a car along a road, maintaining its speed and position at all times as the road bends and sways, while at the same time responding rapidly and accurately to target signs (such as green circles) but not distractor signs (such as green squares). Both of these tasks, driving and targeting, are independently adaptive, meaning that as players get better, the challenge increases (i.e., the car increases its speed and the signs require quicker responses). Another design feature to prevent players from trading off between the two tasks is rewarding them with a "level up" only when both skills improved. *NeuroRacer* thus involves repetitive interruptions and distractions that challenge a player right at the edge of his or her interference resolution ability at all times. Now, we just needed to build it.

So, how do you a build a cognitive control video game if you are an academic laboratory? Pretty straightforward: establish collaborations with video-game professionals. And so, in early 2009 we assembled an all-star team of video-game industry pros (engineers, designers, developers, artists) who agreed to volunteer their valuable time to help build *NeuroRacer*.[33] They worked together with members of the Gazzaley Lab to create this unique interactive training software that was similar to a cognitive exercise in terms of its engine, but with the interface, engagement, rewards, and art of a video game. Under the scientific leadership of Dr. Joaquin Anguera, we then set about performing research on *NeuroRacer* over the next several years to determine if playing it would boost cognitive control in older adults.

In September 2013, we published the results of a series of *NeuroRacer* experiments in *Nature* as the cover story: "Video Game Training Enhances Cognitive Control in Older Adults."[34] We showed that we could use our custom-designed video game as both a cognitive and neural diagnostic tool to understand how multitasking changes throughout one's life. Specifically, we reported that multitasking performance declines from the age of twenty onward and that this decline is accompanied by diminished activity of the prefrontal cortex in older adults at the most challenging moment in the game, when a sign appears and they are also driving. The deficient activity level was of a rhythmic brain oscillation known as *midline frontal theta*, which is generated at the prefrontal cortex and associated with all aspects of cognitive control.

We then performed a video game training study that involved recruiting older adults between sixty and eighty years old who lived in the San Francisco area. Their task was to take a laptop home and play *NeuroRacer* for twelve hours over the course of one month (one-hour sessions three times a week) and then return to the lab so we could see what changed in their performance and in their brains. Some older adults played the multitasking version of the game (both driving the car and recognizing the correct signs) while others played a single-tasking version that exposed them to only one of the two tasks at a time. This study revealed that video game play with the multitasking version of *NeuroRacer* improved multitasking performance of older adults on the game to the level of twenty-year-olds, which remained at this heightened level six months later without them playing again. Their multitasking performance improvement was accompanied by increased levels of midline frontal theta, revealing that the game induced a reversal of age-related deficits in prefrontal cortex activity. In support of our main hypothesis, we also found that the older adults improved in performance on other cognitive control tasks, evidenced by improvements on untrained tests of working memory and sustained attention. *NeuroRacer* even boosted working memory performance when it was assessed in the setting of interference, both distraction and interruption. In terms of the mechanisms of these effects, we showed that it was the multitasking nature of *NeuroRacer* that resulted in these changes, as none of these effects occurred in the older players who engaged in the single-tasking version of the game. Our older participants enjoyed this experience and exposure to new technology, supporting the view that it is a misconception to believe older adults do not enjoy learning how to use modern technology, including video games.[35]

This study taught us many things. Chief among them was that challenging interference processing using adaptive video game mechanics can lead to a transfer of benefits to other cognitive control abilities in older adults. It was also an example to the field of how to carefully design a video game training intervention and perform a well-controlled study. This project propelled the Gazzaley Lab on a path toward developing many novel video games to aid the Distracted Mind.

Although the results of this and other video game training studies are encouraging, we are really just beginning our journey toward elucidating the feasibility and mechanism of this approach. The level of evidence is still a *signal*. We suspect that this will advance to the prescriptive level in the near future, as the combined evidence from cognitive exercises and video game research converge to drive the field to the next stage. Plenty of challenging work still needs to be done to understand the specific design elements and delivery approaches that lead to the most meaningful and sustainable enhancements of cognitive control. Also, not all studies have yielded consistently positive results; these discrepancies need to be better understood.[36] And of course, we need large-scale RCTs and a better understanding of the real-world impact on our Distracted Minds.

It is important to be aware that not all video games are created equal. Some may yield great benefits, others may have no effects on our cognitive abilities at all, and others may even have negative effects. The action video game world has been plagued with claims that they are actually dangerous. Some evidence has supported the claim that violent video games have been associated with a desensitization of players to violence and decreased empathy, although much debate continues over causality.[37] There is also concern about addictive behavior and other negative impacts of video games on real-world behaviors. For example, self-reported attention measures in children and adolescents, notably ADHD-like symptoms of inattention and hyperactivity, were significantly correlated with the amount of video game exposure.[38] This relationship was interpreted as showing that "high excitement and rapid changes of focus that occur in many video games may weaken children's abilities to maintain focus on less exciting tasks (e.g., schoolwork)."[39] However, these data are correlational, so no strong claims can be made as to causality. In either case, it is an important reminder that cognition does not equal behavior. Even if video game technology can be harnessed to boost cognition and alleviate cognitive control limitations that underlie the Distracted Mind, we still may need to influence behavioral factors to attain maximal real-world benefit.

NATURE

In 2007, thirty-eight University of Michigan students, armed with a map and tracked by GPS, took a one-hour walk through either a tree-lined arboretum or a traffic-heavy urban center. Before and after these walks they performed a working memory test. A 2008 paper described a significant improvement in their working memory performance after the nature walk, but not after the urban walk. Similar beneficial effects of nature exposure have been shown to occur in children with ADHD and young adults with depression, and amazingly even in response to just viewing nature pictures.[40] These studies generated evidence that supports the validity of a theoretical construct known as attention restoration theory (ART).[41]

We've already discussed how education, meditation, cognitive exercises, and video game training might boost our cognitive control abilities. The promise is that repetitive exposure to these experiences will harness brain plasticity and lead to enduring brain changes. All of these approaches are active interventions that involve hard work on the part of the participant. ART, however, is an approach that is quite the opposite; it is essentially passive by design. The premise is built on the foundation that top-down, goal-directed, cognitive control fatigues with repetitive use, not dissimilar from how our physical bodies temporarily weaken after we challenge them. This fatigue is associated with subsequent diminished cognitive control, presenting as increased interference on task performance and diminished self-control.[42] This harkens back to our discussion in chapter 5 of the Distracted Mind being a state that is influenced moment to moment by external or internal factors. Cognitive fatigue is one of those factors.

ART proposes that cognitive fatigue can be most effectively and rapidly restored by relaxing the mind from the top-down demands through engagement in a strong bottom-up driven activity, which captures attention not based on goals, but because of the characteristics of the stimuli. The idea is that nature does exactly that; or what Dr. Kaplan, a pioneer of this theory, describes as "soft fascination."[43] Natural environments capture our attention in a bottom-up fashion because natural stimuli are so inherently compelling to us (presumably owing to evolutionary factors). They draw us in but generate minimal top-down responses. This is in comparison to urban environments where bottom-up stimuli are more likely to induce a cascade of top-down activities. And so, the bottom-up journey of a nature hike that the study participants engaged in was hypothesized to have given their top-down cognitive control a break, and time needed to restore their cognitive control resources and improve their working memory. Others have suggested that stress-relieving benefits of being immersed in nature also contribute to restorative effects.[44]

Despite a steady accumulation of evidence supporting this theory, the field is still very young and the current evidence places it in our view at the level of a *reasonable hypothesis*. We need more objective documentation of cognitive fatigue recovery being directly caused by exposure to nature, as well as a better understanding of underlying mechanisms to advance this approach to the next level.[45] That being said, exposure to nature is highly encouraged for its many benefits on our health and mind.

DRUGS

If exposure to nature takes us a step closer to a more passive approach to aid the Distracted Mind, we might be tempted to ask about approaches that go even further in that direction. For those who truly want to take the easy route to building a stronger brain, the most passive option is to simply pop a pill. Our increasing understanding of the role of neurotransmitters and other neuromolecules in cognitive control has expanded the possibility of drugs being used to boost these abilities and diminish our limitations. Indeed, many substances have already made such claims. They fall under broad classifications as "smart drugs," "cognitive enhancers," and "nootropics," which are often pharmaceuticals used for therapeutic treatment of clinical conditions such as ADHD, narcolepsy, and Alzheimer's disease. But drugs like amphetamine, methylphenidate, modafinil, and cholinesterase inhibitors have also made their way into the mouths of healthy individuals who are interested in a cognitive boost.[46] In fact, their usage on college campuses for this purpose has been reported as being as high as 25 percent.[47] This raises the critical question: do they actually increase cognitive control?

The stimulants, methylphenidate (e.g., Ritalin) and amphetamine (e.g., Adderall), are the most widely used drugs to treat ADHD, and also the most widely used for nonmedical purposes by healthy individuals. They act to increase levels of norepinephrine and dopamine throughout the brain, while their cognitive effects have largely been attributed to their actions at the prefrontal cortex.[48] Although these drugs have clear clinical benefits for children diagnosed with ADHD, it seems these are largely behavioral changes reflected in improved classroom management, rather than an actual remediation of cognitive control deficits or improved academic performance.[49] When it comes to cognitive control, both clinical and healthy populations seem to show some benefits in working memory and sustained attention on simple tasks, but on more complex tasks that demand greater control, such as those that involve interference, the results are less clear.[50] In fact, sometimes these drugs may even impair cognition under these circumstances.[51] Different response effects across dosage levels (e.g., cognitive effects are largely confined to lower doses), as well as significant individual differences further complicate this story, leave it unclear as to whether these drugs are truly cognitive control enhancers. This seems to be in stark contrast with the widespread impression that drives their rampant use on college campuses.

Another drug that has gained a lot of interest in this domain is modafinil, which is clinically used as a wakefulness-promoting medication for medical conditions such as narcolepsy. Just like the stimulants, modafinil has become widely used as a cognitive enhancement drug that has joined the ranks of prescription drugs adopted by healthy individuals for recreational purposes.[52] Studies have revealed that modafinil does seem to enhance aspects of cognitive control in well-rested, healthy individuals. This has been documented as improvements in sustained attention, selective attention, and aspects of working memory, with the effects boosted in the setting of sleep deprivation.[53] However, there are many inconsistencies across the results of different studies in the extent to which this drug enhances cognitive control, highlighting the complexity of influencing such a complex system with a single molecule.[54]

It seems that the level of evidence is that there is a *signal* that some pharmaceuticals enhance aspects of cognitive control, although the effects appear to be modest in magnitude. The reality is that there is currently no magic drug for the Distracted Mind. This is especially true since molecular approaches generate very real concerns about side effects and addiction. There are also ethical issues at the societal level regarding "brain doping" with an intervention that can be performed so rapidly in an effortless manner, in terms of a loss of authenticity, inequality, coercion, and cheating.[55]

PHYSICAL EXERCISE

Let's swing the pendulum all the way back in the direction of effort-demanding approaches to aid the Distracted Mind and discuss the most "active" of them all: physical exercise. We are sure most of you are now well aware that physical activity, and of course the more formal practice of physical exercise, has well-documented benefits on human health, including cardiovascular disorders, cancer, obesity, diabetes, and stroke.[56] But these benefits also extend to mental health with evidence of a positive impact on symptoms of anxiety, depression, and schizophrenia.[57] These findings have been complemented by an onslaught of fascinating data regarding neural changes induced by exercise, which span the gamut from increases in brain volume (both gray and white matter), nerve growth factors, blood flow, functional and structural connections, and even new neurons being born.[58] Perhaps not surprisingly, this neural plasticity is accompanied by a host of cognitive benefits, a claim that has been supported by several meta-analysis studies.[59]

When it comes specifically to the impact that physical exercise has on cognitive control, the last two decades have been host to many studies on this topic in both children and young adults. The higher-level control abilities of a preadolescent child who is more physically fit have been shown with both performance data and neural recordings.[60] The general conclusion is that "lower-fit children may have more difficulty than higher-fit children in the flexible modulation of cognitive control processes to meet task demands."[61] This same relationship has been shown to exist for college-age adults.[62]

In terms of a real-world impact of physical fitness on the Distracted Mind, one study compared the ability of lower-fit and higher-fit children to navigate a heavily trafficked road by having them walk on a treadmill in a virtual reality environment. The trick was that they had to navigate while walking undistracted, listening to music on an iPod, or talking on a hands-free mobile phone. Previous research has shown that mobile phone use impairs the safety of actual street crossing by pedestrians of all ages, but children are particularly susceptible to being hit by cars while distracted by their phones.[63] This study found that higher-fit children were more successful in street crossing under all testing conditions. Moreover, higher-fit children were not negatively impacted by the phone and music distractions, while the lower-fit children showed worsened performance when either listening to music or talking on the phone compared to undistracted crossing.[64] These data are consistent with the results of laboratory cognitive tasks and support the conclusion that being more physically fit when young means having a less Distracted Mind.

An even more powerful approach is to generate causal data using an interventional exercise design. These studies have yielded results that are consistent with the high-fit versus lower-fit findings. They have shown that aerobic exercise training for children results in improvements on cognitive control tasks.[65] Boosts in cognitive control abilities occur even after engagement in a single bout of physical exertion, as assessed in healthy children and those diagnosed with ADHD, with benefits extending to academic achievement.[66] Interestingly, it seems that the impact on the brain is greater if an exercise program is also cognitively engaging. Similar training benefits of acute and chronic exercise on cognitive control have been shown in both young adults and middle-age adults.[67]

There is also a very large body of research on the cognitive benefits of physical exercise in older adults. A landmark meta-analysis of exercise interventions published in 2003 showed that there was an overall cognitive benefit of physical exercise in older adults, with the largest effects being in the domain of cognitive control.[68] Interestingly, they also found that positive results were greater for groups engaging in combined strength and aerobic training, compared to those who did aerobic training alone. Although some studies have yielded less convincing results, the general finding of improved cognitive control in older adults has been supported by a more recent review paper and another meta-analysis.[69] They both show benefits for older adults across all aspects of cognitive control: working memory, attention, and goal management.[70]

In terms of the neural mechanisms of these effects, an fMRI study revealed greater activation in the prefrontal cortex in high-fit versus lower-fit older adults while they were performing an interference challenge. Importantly, the same results were found in an intervention study with a group of older adults who trained aerobically for six months versus a control group who engaged in stretching and toning (a similar result was found for overweight children).[71] The study showed not only an increase in brain activity in the prefrontal cortex, but also a decrease in impact of distraction on task performance.

With all of this evidence, we feel that engagement in a physical exercise program should be considered a *prescriptive* approach to ameliorating a Distracted Mind, especially for children and older adults. There exist positive meta-analyses of randomized controlled studies for individuals across the life span, it is low cost and widely accessible, it is associated with many other health benefits, and there is a strong biological basis for its effects. Of course, plenty of work still needs to be done, such as determining the ideal length of an intervention, duration of each session, the specific elements of training, real-world impact, and interactions with other approaches. For example, nutritional programs and combined therapy with cognitive exercises might enhance the effects.[72]

NEUROFEEDBACK

In 2010, fourteen students enrolled in a research study in Germany that involved reporting to a laboratory each day for five days.[73] There they would sit in an electrically shielded, soundproof room wearing an EEG cap while staring intently at a computer screen displaying a simple square. Their goal was to turn the color of the square from gray to red using nothing but their minds. They were not given explicit instructions on the best strategy by which to accomplish this seemingly bizarre task. But they did know that their brain activity was being recorded and that it was being used to change the color of the square. They also knew that the redness would become more saturated when they concentrated in such a way that increased a specific brain activity rhythm, and that it would turn blue and become a more saturated blue if they concentrated in the wrong way. Using this approach, known as "neurofeedback training," the participants gradually learned through trial and error how to turn the square redder and redder, with most of them reporting "evoking emotions" as the best strategy. Since the redness of the square was directly associated with their brain activity—in this study it was the alpha rhythm—over the course of one week, eleven out of the fourteen students were successful in increasing their alpha rhythms gradually each day. Most interestingly, by the end of the week they also showed improvements on a cognitive control challenge of mental rotation compared to a control group. A more recent study showed that neurofeedback training of alpha rhythms resulted in changes in neural networks even after a single thirty-minute training session, with the degree of network change correlated with a reduction in mind wandering on an attention task.[74]

The rationale for neurofeedback, which began in the 1960s, is that since neural rhythms underlie many aspects of cognition, if you can learn to enhance a specific rhythm through feedback training, then there will be benefits to cognitive skills that depended on those underlying brain rhythms.[75] This approach is a form of brain–computer interface (BCI) that allows an individual to gain voluntary control over brain rhythms by receiving real-time, reward-based feedback on how activity is being modulated when he or she thinks in a specific manner. The neural recordings are usually done with EEG, although it has been successfully accomplished using other techniques, such as fMRI and NIRS. The feedback is performed using visual (and sometimes auditory) representations of the magnitude of a brain rhythm. This often takes place in the setting of a simple game that gives moment-by-moment indications of success and failure of the individual's ability to move a specific brain rhythm(s) in a specific direction.

Clinical applications have been the primary focus of the neurofeedback field, with some, albeit limited, success in ADHD, autism, anxiety, addiction, and depression.[76] More recently, interest has emerged in the use of neurofeedback to optimize cognitive performance in healthy individuals. A recent multisession neurofeedback study directed this training approach at increasing activity of midline frontal theta (the same rhythmic brain activity measure that we enhanced in older adults with *NeuroRacer*). The researchers showed that midline frontal theta neurofeedback resulted in an increase in activity on untrained cognitive tasks and improvement in working memory and task-switching performance. Another study revealed that a similar approach improved working memory and selective attention performance in older adults.[77] The mechanisms of neurofeedback effects are just now being elucidated, although it seems that in addition to neural activity changes, structural brain changes are induced in both the gray matter and white matter.[78] Although neurofeedback as a tool for enhancement is not ready for prime-time use, a steady accumulation of studies suggests the presence of a *signal.* It would seem that some version of this approach might indeed be useful in strengthening cognitive control abilities.

BRAIN STIMULATION

The final approach we will discuss that might give our Distracted Mind a boost is direct stimulation of the brain with electrical and magnetic fields. Brain stimulation is based on the premise that since the brain's functioning relies on electrical signaling we can influence its function by externally (or internally) applying electrical and magnetic fields. While this may be the most science fiction sounding of all the methods we have covered, it has been around for over a hundred years and is becoming more prevalent. Even though some scientists feel that it should still be firmly planted in the realm of the research world, we already see companies selling consumer brain-stimulation devices for this purpose. An approach that has been considered for over a century has finally come of age.

While there are many forms of electromagnetic brain stimulation (e.g., alternating current, strong and weak magnetic fields, and random electrical noise) the most widely used approach for cognitive enhancement is transcranial direct current stimulation, or tDCS.[79] This brain stimulation approach exploded onto the scene after a study in Germany by Dr. Nitsche and Dr. Paulus in 2000 showed convincing evidence that weak currents could modify the neural responsiveness of the cortex.[80] In most tDCS studies that have taken place since then, participants have received just a few milliamps of current directly to their scalp (a 100-watt lightbulb draws 500 times that level of amperage), often delivered using a simple nine-volt, battery-powered device. Depending on how the electrodes are placed, it will either amplify or suppress the likelihood of neural firing in the underlying brain regions. Remarkably, this has been shown to have lasting effects after the current is turned off. Experiments are then performed to assess the influence of this stimulation on learning and cognition.

Over the years, researchers have studied tDCS as a treatment for neurological and psychiatric conditions, such as depression, Parkinson's disease, and stroke, with some convincing successes.[81] If future research continues to generate positive results with low side effects, we are likely looking at a whole new toolkit of therapeutics—what we may come to call "electroceuticals." In addition to clinical applications, there has been great interest in tDCS as a cognitive enhancement tool to optimize brain function in healthy people.[82] There is now evidence that tDCS (as well as other types of brain stimulation) can have a positive impact on the Distracted Mind by boosting core cognitive control abilities of attention, working memory, and goal management.[83]

For example, a recent study assessed the impact of tDCS applied to the prefrontal cortex on the sustained attention abilities of nineteen military personnel who volunteered to receive brain stimulation while they were engaged in a simulated air traffic controller task that involved identifying rare targets over a forty-minute period.[84] The data showed that, compared to a control group whose performance declined as expected over time, the tDCS group showed significantly higher levels of vigilance by largely retaining their target-detection performance over the entire forty-minute session. The authors interpreted these results as indicating that "tDCS may be well-suited to mitigate performance degradation in work settings requiring sustained attention."[85] Extending this field to the domain of interference resolution, the Gazzaley Lab recently explored the influence of tDCS applied to the prefrontal cortex on multitasking abilities with our video game *Neuro-Racer*.[86] We found that stimulation resulted in performance improvements after a single session, but only when multitasking, not for the single-tasking version of the game.

There is now growing interest in using transcranial alternating current stimulation (tACS) as a tool to boost cognitive control abilities. Unlike direct current (DC) stimulation, the use of alternating currents (AC) opens up the potential to directly target stimulation to different brain frequencies, with the hope that this will result in more selective effects. The basic idea is that the frequency of the applied alternating current might amplify those same naturally occurring frequencies in the brain that underlie cognitive control abilities.[87] Early studies have started to show the promise of this approach to increase the power of brain rhythms after the stimulation is turned off and to boost cognitive control, for example, working memory capacity.[88] Much more research is needed to confirm these effects and understand the mechanisms.

In general, there are many reasons to be excited about electrical brain stimulation as a promising tool to enhance our cognitive control abilities and benefit the Distracted Mind, but it is currently at the level of *signal.* Many critical questions still remain to be answered before this is a viable and prescriptive approach—not the least of which is having a better understanding of long-term side effects.[89] In addition, there may be negative consequences, such as the improvement of some aspects of cognition at the cost of others.[90] Ethical issues surrounding impact on personal identity, autonomy, and authenticity should be addressed before we are advance into widespread consumer use of a new technology that could dramatically boost brain function.[91]

CONCLUSIONS

There is quite a wide range in the strength of evidence that has been generated for the many approaches we have considered to aid the Distracted Mind. In our opinion, only physical exercise has reached the highest level of being a truly prescriptive approach. Not far behind this are cognitive exercises, video game training, and meditation, where we see a strong signal in the accumulated scientific literature. It will take more time to work out the details of exactly what is the best formula and delivery system of these approaches. But we are cautiously optimistic that they will have an important role in the future. There is a weaker signal, but a signal nonetheless, for other approaches that need more evidence to become prescriptive—drugs, brain stimulation, and neurofeedback. Both traditional education and nature exposure are still at the level of being reasonable hypotheses, but they deserve further research. We find that innovative approaches in the domain of education have already started to accumulate evidence (e.g., *Tools of the Mind*). Also, it is important to note that these are broad categories, with many possible instantiations. So, although there may be a strong signal that an approach, such as cognitive exercises, may work, it does not mean that all exercises will be effective.

There is also increasing interest by scientists to study how these approaches can work together to produce synergistic effects, such that the sum is greater than the parts.[92] This is known as a "multimodal approach," or as we like to call it in the Gazzaley Lab, "neuro cross-fit training." The Gazzaley Lab has already started seeing early evidence of beneficial effects in combining physical and cognitive exercise.[93] The lab has now created a novel video game, *Body-Brain Trainer*, as an integrated physical and cognitive fitness challenge that uses the *Xbox Kinect* motion capture game system. We will also incorporate meditation principles and nature exposure into game play. We are optimistic that the future will reveal powerful synergistic interactions between approaches can be customizable to individual differences and lead to meaningful and sustainable improvements in the brain.

Even if cognitive control abilities can be increased via these approaches, it seems unlikely that any or even a combination of approaches will eliminate all manifestations of the Distracted Mind. When we eventually have a neuro cross-fit training program that minimizes cognitive control limitations in our brain, there is still no guarantee that this will translate into a positive impact on everyday life activities. This is because, while cognitive control is essential for all higher-level interactions, there is no one-to-one mapping between cognition and real-world behavior. Environmental factors may dominate that do not allow the benefits of enhanced cognitive control to have an optimal impact on daily life. For example, technology-induced anxiety may mask benefits of having superior attention abilities. The ideal approach, it would seem, is to change both your brain and your behavior. In the final chapter of the book, we will offer advice from a practical perspective on how to modify your behavior with strategies to optimize the performance of your Distracted Mind.

11 MODIFYING BEHAVIOR

IT SHOULD NOW be abundantly clear that we live amidst a level of high-tech interference that in the past decade or so has dramatically changed the world, and along with it our thoughts, feelings, and behaviors. In Part II we explored the many ways in which modern technology has aggravated our Distracted Mind; from awakening in the morning until trying to fall asleep at night, we are tempted by technological distractions and interruptions. As we have shown, three main game changers—the Internet, smartphones, and social media—have forever altered our mental landscape. We have painted a detailed picture based on solid research from a variety of fields showing that we are spending our days switching from task to task and affording each only our divided attention.

Recall the cognitive control limitations that we presented in chapter 5 in the domains of attention (*selectivity, distribution, sustainability, processing speed*), working memory (*capacity, fidelity*), and goal management (*multitasking, task switching*). As described, high-tech influences stress these limitations in just about every possible way: they challenge our attention abilities via frequent distractions, fragment our working memory and diminish its fidelity through interruptions, and drive us to excessive multitasking and task switching, all of which introduce performance costs. In terms of the MVT model, introduced in Part I and elucidated in chapter 9, modern technology has caused this by diminishing the time in which we remain engaged with an information source, causing us to shift to another patch before we have exhausted the information in our current source. We are like a squirrel with an attention disorder, constantly jumping from tree to tree, sampling a few tasty morsels and leaving many more behind as he jumps to the next tree, and the next and the next. It sounds exhausting, and, as we have shown, it is negatively affecting our safety, relationships, school and job performance, and mental health.

It is time to take control. In the previous chapter we presented ways to harness plasticity and enhance brain function to boost cognitive control abilities, allowing you to be more resilient to distraction and multitask more effectively when needed. Here we propose a complementary and parallel approach to your efforts in boosting cognitive control. We will present strategies by which you can modify your environment and your behavior to better focus on a single task and ease the strain on your cognitive control limitations, thus diminishing the challenges on your Distracted Mind and improving your performance and quality of life.

Before we introduce these strategies, let's consider when and why we would want to do this. Do you always need to stay laser focused on a single task? Is there any time that you might actually want to allow yourself to be exposed to distractions and interruptions? We learned in chapter 4 that task switching carries a cost and trying to do two things at the same time means that you can't give your full attention to either. But surfing the web or texting with multiple friends can be fun. One study that looked at the impact of interference on boredom established that external interruptions (alerts, notifications) reduced boredom on simple tasks, but if the task was complex, or even if it was simple but required sustained attention, external interruptions actually increased task-related boredom.[1] We recommend that you think about the task in front of you and decide whether it needs your undivided attention. When you have simple, noncritical tasks that are not time-sensitive and do not demand sustained attention, then by all means consider keeping your phone on and browser windows open. Having a more enjoyable time multitasking in this situation may actually be what allows you to accomplish a set of low-priority tasks that really just need to get done.

But if you need your brain to be at peak performance, then your goal should be sustained, focused, and singular attention. A good rule of thumb for when you should minimize switching and give a task your full attention is if it is: (1) difficult or requires a lot of thought (e.g., studying for a

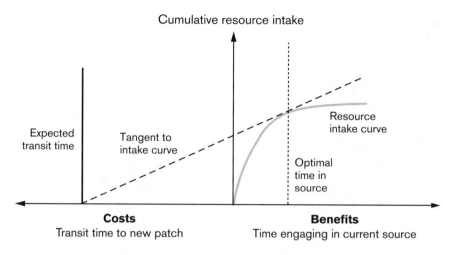

Cumulative resource intake

Expected transit time

Tangent to intake curve

Resource intake curve

Optimal time in source

Costs
Transit time to new patch

Benefits
Time engaging in current source

Figure 11.1

A graphical representation of the marginal value theorem.

challenging exam); (2) carries a risk of high negative impact (e.g., driving a car); (3) is critical or high value (e.g., writing a work proposal or engaging with your loved ones); or (4) is time-sensitive (e.g., preparing a report for work that is due at the end of the day).

We have turned to the MVT model throughout this book, initially to explain why we behave in this manner in the first place, and then to explain how technology has influenced our behavior and exacerbated our Distracted Minds. The goal in this chapter is to use the MVT as a framework to present practical strategies to avoid distractions and interruptions, so that we can optimize the productive time we spend in an information patch.

Let's begin our discussion of strategic approaches to modifying behavior by reviewing the MVT model (figure 11.1). The optimal time spent at an information source (i.e., how long you stay focused on the current task) is based on the "expected transit time," shown on the left side of the model, and the "resource intake curve," shown on the right. The longer the expected transit time (a move to the left reflects a greater expected cost of switching) and the larger the resource intake curve (a move up the graph to the right reflects a greater benefit of staying), the longer you will remain at your current source. Both the transit time and resource curve are affected not only

a.

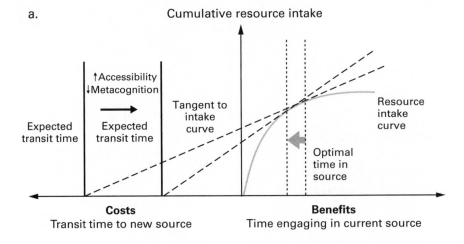

b.

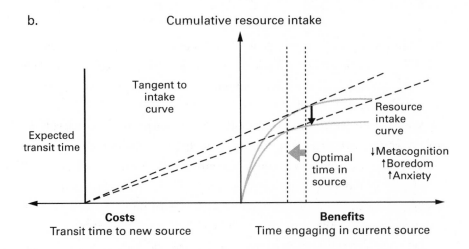

Figure 11.2
A summary of the influences of information technology on the marginal value theorem factors that result in more frequent multitasking and task switching.

by external factors, but also by our subconscious estimates. This means that we can modify these elements of the model either by changing external factors or by changing our own thoughts and therefore our internal estimates.

Chapter 9 described how left-side aspects of the model—the expected transit time to new patches (figure 11.2a)—have been affected by increased accessibility to new information sources via the wonders of modern high-tech (external), coupled with a lack of metacognition resulting in us underestimating our multitasking and switching costs (internal). These both lead to decreased expected transit times (a move to the right), and, in turn, result in diminished time spent at an information source. Similarly, technology affects the right side of the model—flattening the slope of the resource intake (figure 11.2b)—via an increased rate of boredom and anxiety accumulation when engaged in a source, as well as a metacognitive undervaluing of the benefits of remaining in a current source (internal), again decreasing the time spent with a source.

To manage these high-tech influences that challenge our cognitive control limitations and exacerbate our Distracted Minds, we propose that you can maintain better focus on an information source by taking the following steps (figure 11.3):

1. Improve *metacognition* by increasing your understanding of the cost of multitasking/task switching (left side) and the value of remaining at an information patch (right side).
   - One of the main objectives of this book has been to accomplish this very goal by sharing a wealth of information on the underlying basis of our Distracted Mind.
2. Limit *accessibility* to new information sources (left side).
   - We'll outline some simple strategies to keep these temptations a little further away.
3. Decrease your *boredom* when focusing on a single goal (right side).
   - We'll discuss approaches to make engagement in tasks more fun and productive without jeopardizing the primary goal.
4. Reduce *anxiety* that prompts a switch to something new (right side).
   - We'll describe actions and technologies that help prevent FOMO from a social perspective.

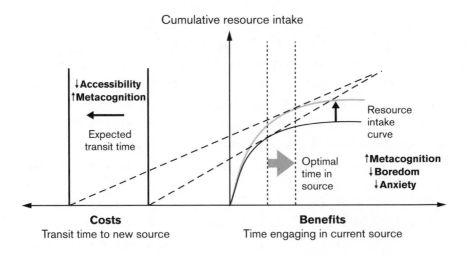

Figures 11.3

By modifying several factors—increasing metacognition, decreasing accessibility, diminishing boredom, and reducing anxiety—we can diminish counterproductive behavior.

Let's tie all of this back to the MVT model: (1) (↑ metacognition) will shift the expected transit time toward the left and increase the resource intake curve; (2) (↓ accessibility) will drive the expected transit time to the left; and (3, 4) (↓ boredom and anxiety) will increase the slope of the resource intake curve (figure 11.3). All of these will influence factors of the MVT model that increase the time you remain at an information source, thus helping you to manage this interference dilemma.

Before we explore strategies on how to change these factors, it is important to note that there is another school of thought on how to deal with technology. If you treat an obsession with technology as an "addiction," then a natural "cure" would likely be similar to how many treat a drug or alcohol problem: with a detoxification program. In just the last few years we have seen books proclaim that the way to regaining our focus and attention is through a "digital detox" program. We are urged to unplug, to take a holiday from technology, and even encouraged to find JOMO or the "joy of missing out." The popular press regales us with detox success stories ranging from one man living with no Internet for a full year, to learning how to earn

"99 days of freedom," to shutting down your Facebook page (and saving twenty-eight hours of lost time spent immersed in social media), to "Camp Grounded" where you spend a weekend retreat with no technology, to the "Unplugged Project" that encourages us to learn what it is like to go twenty-four hours without technology, to a national day of unplugging, and finally, to evening technology-free parties in bars—where people realize that even a few hours without their technology makes them anxious and leaves them feeling disconnected, even as they are connecting to living and breathing human beings sharing the same physical space.[2]

While taking a break is certainly a good thing and can act to improve our metacognition about the influence of high-tech on our minds, to put it simply, there is no evidence that *extended* tech detoxes actually work.[3] Sure, you might feel better for an evening, a day, or even a weekend, but when that detox time is over you are right back to your information-foraging behavior, frantically dividing your attention to catch up with all that you missed out on while technologically disconnected. Just like short-term diets and drug or alcohol detox programs, unless you work to change your environment and routines, it won't be long before you return to the same old habits. In terms of the MVT model, information that was not available during your detox is now available in much larger quantities, which then drives even more frequent interruptions and in the end, a more Distracted Mind.

We have talked to many people who attempted to take extended breaks from technology. A person who is detoxing initially reports feeling a sense of anxiety for all that he or she is missing out on, which seems to abate over time. However, upon returning to their tech-rich world, detoxers find themselves even more buried in the text messages, phone calls, emails, and social media posts that they missed. Given that the average teen sends and receives about 3,200 texts a day, that the average adult gets hundreds of email messages a day, and that most people's social media feeds are bustling with dozens of daily posts, comments, and photos, returning to the "real" world after detoxing means hours of getting back to any semblance of a state of normalcy. And "normal" means that even while "catching up" on missed connections, new ones are arriving, often punctuated by insistent alerts and notifications, all reminders that taking an extended break means more work and ultimately more stress. In place of ineffective technology detoxes we offer you a variety of strategies to help your Distracted Mind. This will provide you with a path to taking back control, so that you are the one who decides when you are ready to accept an interruption and when you need to ignore that interruption and maintain focus on what you're currently doing. We will present these strategies for reducing goal interference by considering four common scenarios.

Scenario 1: Driving a Car

Even though most of us know that task switching and distraction while driving can lead to an accident and possible injury or even death, many continue to engage in this behavior. You see people stopped at a light, face down, clearly looking at their phone and when the light changes it takes a horn honk to get them moving again. You watch the driver next to you on the freeway glancing down and up again and then back down again, clearly looking at a smartphone on his lap. You might even sneak a quick peek at your own phone thinking that you're a good driver so it should be fine.

Scenario 2: Focusing on Critical Assignments

Almost everyone faces a critical assignment at one time or another, from children working on school projects to professionals drafting proposals to academics writing papers to musicians editing new tracks. Even though it is now fairly well known that task switching and multitasking reduce productivity, distractions are everywhere: classrooms are full of devices and open offices have become the norm. Anytime you sit down at a computer, a plethora of interruptions is just a click away, if not already bouncing up and down begging for your attention.

Scenario 3: Social Interactions

Whether we're having mealtime with the family, interacting with colleagues at work, or sharing dinner with friends and family, technology is putting pressure on our relationships. The new expectation is that we will respond instantly to a text message even when that means ignoring the person sitting in front of us. Just take a look around and you'll see devices everywhere, weakening our interpersonal connections and threatening to interrupt us at any moment. New research continues to discover more reasons why face-to-face interaction without the presence of technology is critical for maintaining healthy relationships.[4]

Scenario 4: Sleep

Chapter 5 outlined the impact of sleep deprivation on cognitive control, chapter 6 showed how our technology is affecting our sleep, and chapter 7 discussed many other negative consequences of lack of sleep. We need to get more sleep, but our busy lives, coupled with high-tech influences, push us to spend less time in bed and more time staring at our screens. This can become a vicious cycle of sleep deprivation, impairing our judgment and making us less productive, followed by us sleeping even less and attempting to do more things at the same time to catch up.

We will now turn to each of these scenarios and share practical approaches that will serve to increase our metacognition, decrease the accessibility of technology, decrease our feelings of boredom, and minimize our anxiety of missing out, thus putting pressure on factors that live on both sides of the MVT model to increase our optimal time in an information source when it is most important.

STRATEGIES FOR SAFE DRIVING

In spite of costly fines and well-publicized dangers of using a phone while driving, many people cannot resist the clarion call of an incoming message. Now we have a dangerous situation where our phones are right next to us and can be accessed via the car's built-in Bluetooth (high accessibility). In addition, many believe that because they are good drivers they are able to multitask while on the road (poor metacognition). This is made worse when the route is very repetitive (boring), and is further aggravated by pressure to be reachable even while driving (anxiety). Put this all together, and it is a recipe for disaster that plays out all over the world every day. Here are some strategies that may help us ignore the lure of the phone and learn to delay a need to check in.

↑ Metacognition

Hopefully, the content presented in Parts I and II helped you to truly understand not just how dangerous driving while distracted and multitasking is—that "text messaging creates a crash risk 23 times worse than driving while not distracted,"[5] that using a cell phone increases your chance of being in an accident as much as being legally drunk—but also *why* we are so vulnerable to it, and why it's so hard for us to stop. If you still think you're a master of multitasking or just want to experience for yourself how detrimental texting and driving is, there are several online simulators where you can take a spin.[6] Virtually experiencing the severe consequences of texting and driving may be enough to help us change our metacognition about our abilities.

The research on texting and driving is so clear that all but six states have banned it.[7] In contrast, while fourteen states have banned using handheld phones for all drivers, none have included a ban on all cell phone calls despite research demonstrating that handsfree cell phone use also increases the risk of accidents.[8] This illustrates the pervasiveness of our poor metacognition. The bottom line is that as soon as drivers remove their attention from the road to do anything other than drive it induces risk, and we all need to be very aware of this.

↓ Accessibility

The simplest way to remove the temptation to text or talk on your mobile phone while driving is to put your phone in the trunk (with Bluetooth off). There are also several apps available now that block texting and making phone calls while driving. Some, such as DriveOFF and DriveMode, are based on speed: once you are traveling at more than a slow speed, the app blocks your ability to text, displays a static screen saver, or reminds you not to text (which may be distracting, in and of itself). Additional apps are available to curtail cell phone distractions and interruptions including Live2Txt, TxtShield, SafeDrive, and LifeSaver. In addition, if you are a parent of a driving child—and we know how scary that can be—there are apps that will monitor tech activities while your teen is driving including Canary and DriveSafeMode. And by the time this book is in print there will surely be dozens more apps, so consult your app store. If this is too extreme for you, give your phone to a passenger when you are driving and let him or her respond if needed.

↓ Boredom

There are ways to make your regular commute more fun and engaging without putting yourself at risk. For example, talking to a live passenger does not have the same negative impact as talking on the phone, although according to recent research this depends on the mental demands inherent in the conversation.[9] The more your conversation requires cognitive control resources, the more talking with a passenger may lead to impaired driving. If you need to have a meeting while driving, try carpooling and have your fellow commuter drive while you devote full attention to your meeting. Noninteractive activities are also much safer than phone calls, texting, or emailing. In a recent study, David Strayer developed a "Workload Rating Scale" and found that compared to speaking on the phone, listening to an audio book or listening to music required reduced workload and thus presented less chance of an accident.[10] Given that audio books and podcasts do not appear to require extensive attentional resources, they may actually be beneficial by allowing you to listen to valuable information, in a manner you can control. If driving conditions warrant more of your attentional resources, you should turn off the podcast or audiobook, and return your focus to the task at hand. Finally, even just varying your route can make the drive more interesting and enjoyable.

↓ Anxiety

As discussed in chapter 9, many of us have a lot of anxiety around either just missing out in general (e.g., on a social media post, a text message, or something else in your virtual world), or missing something actually important, perhaps even an emergency. Here are some strategies that can help reduce this anxiety, by letting you take control of how others contact you. First, set clear expectations with family, friends, and colleagues. In the case of driving, let them know your normal commute schedule and that you will not be available during that time. Second, ensure everyone knows of your new plan; you can set up auto-responses to texts and phone calls using various apps such as Live2Txt. These can let callers know that you are driving and will get back to them once you arrive. Third, there are many apps and phone settings that will allow emergency calls through but will not disturb you with less important notifications. Some alert you only if the same number has called multiple times. Others only allow calls from specific numbers. Of course, if you are driving and receive a critical call, pull over to handle the call.

STRATEGIES FOR COMPLETING CRITICAL ASSIGNMENTS

Whether working on a critical report at the office or a school assignment due the next day, we are all faced with a constant stream of distractions and interruptions. In the office we often juggle our work assignments between messages from those pink "While You Were Out" slips sitting atop our desk or responding to urgent email requests while trying to craft an important report. We believe that we'll be able to manage everything at the same time, and maybe even more productively (poor metacognition). A barrage of interrupting email messages, texts, Snapchats, and notifications of social media posts (accessibility) equally beckon a student studying at home to switch attention from the less interesting (boring) work at hand. Add to that the research we have presented showing that a large portion of our interruptions come not from outside alerts but from internal pulls to check in with our virtual world (anxiety), and it is no wonder why we all struggle to stay focused. Here are some strategies to use when you are faced with a critical assignment and an environment that constantly interrupts your thoughts.

↑ Metacognition

By now, you should appreciate how the limitations of our Distracted Mind impact our performance on critical assignments. In chapter 7, we saw this impact on students: multitasking while studying predicted a lower GPA, using technology in the classroom led to lower test scores and productivity across all grade levels ranging from elementary school to college, and engaging in interfering technology in the classroom was associated with an increase in high-risk behaviors for college students. In the workplace a seemingly brief interruption can lead to nearly half an hour off task. Disrupted work may be completed faster but at the cost of a higher workload, more stress, higher frustration, more time pressure, and increased effort. It is important for you to truly appreciate how much time you are actually spending on different online or smartphone activities. Try TrackTime or RescueTime for your computer; and Checky, Moment, Instant, or Menthal will alert you to your daily smartphone use.

↓ Accessibility

A major problem in completing critical assignments, especially on a computer, is the constant availability of that most sought-after commodity: information. Here are some suggestions to help you reduce the accessibility. Begin by setting up your work environment to avoid being distracted and interrupted. *This is the most difficult and challenging part of the process.* Limit yourself to a single screen.[11] Yes, multiple screens are nice for spreading out your work, but they create distractions. In addition, put away all nonessential work materials on your desk, leaving only paper materials that you absolutely need to complete a task. When that is done, get rid of those distracting books and notes. Whenever possible simply find a quiet environment devoid of other people and the presence of interruptions. If you must work in a noisy environment such as a coffee shop, consider wearing noise-canceling headphones. If you are on an airplane flight, consider using those headphones, and also make a conscious decision as to whether you are going to get Internet access.

The next step is to decide which programs or apps you are going to *need* open to complete the task and *close down all others*. Don't just minimize them: really shut them down. Minimized icons beckon to you to open them, and that draws your attention, if only momentarily, away from your task. If you need to access websites to perform your work, open only one at a time. Whenever possible, do not use tabs; and when you are done with a website, shut it down it rather than minimizing or keeping it in your browser. Open but minimized apps and tabs are more accessible than those that are closed, even if they appear only trivially so, and thus they facilitate switching.

Because of its ubiquity, email has proven to be a special case in aggravating a Distracted Mind. You may find it difficult to shut email down, but it is essential that you do so to remove the temptation to respond to the "ding" alert of an incoming message. As we now know, it can take you up to twenty or thirty minutes to return to your work once you allow an interruption. In an interesting *New York Times* article, Clive Thompson, author of *Smarter Than You Think: How Technology in Changing Our Minds for the Better*, argued that we need to end the "tyranny of 24/7 email."[12] Thompson provides examples of major companies such as Daimler, Volkswagen, and Deutsche Telkom who have adopted policies that place clear limits on after-hours email. Edelman, a global public relations firm based in Toronto, has a 7-to-7 rule that discourages employees from sending email messages before 7 a.m. or after 7 p.m. This is not an unreasonable personal strategy to adopt to decrease the accessibility of email to your brain.

What else can you do to end the 24/7 tyranny of email? A recent study by Kostadin Kushlev and Elizabeth Dunn of the University of British Columbia provides a starting point.[13] The researchers divided 124 adults into two groups. For the first week of a two-week study, Group 1 was told to check their email as often as they could while Group 2 was told to just check their email three times a day. For the second week, the groups switched strategies, and Group 1 was asked to check email only three times a day while Group 2 was directed to check as often as they could. Results indicated that when participants—a mixture of college students and community adults—checked only three times a day they reported less stress, which predicted better overall well-being on a range of psychological and physical dimensions.

Based on findings that people are checking their email (and text messages and social media) all day long plus the results of this study that show limiting access to email provides emotional and physical benefits, you can follow the advice of author and McGill University professor Daniel Levitan, who urges people to check electronic communications only at certain times of the day.[14] Levitan advises readers that

> If you want to be more productive and creative, and to have more energy, the science dictates that you should partition your day into project periods. Your social networking should be done during a designated time, not as constant interruptions to your day. Email, too, should be done at designated times. An email that you know is sitting there, unread, may sap attentional resources as your brain keeps thinking about it, distracting you from what you're doing.

What might be in it? Who's it from? Is it good news or bad news? It's better to leave your email program off than to hear that constant ping and know that you're ignoring messages.

Once you have limited the accessibility of interrupting computer screens and programs, notably communication treadmills like email and messages, you should silence your smartphone. Turn off all alerts of any kind, including vibrations, and, if you still feel the pull, move it to another room. If you need the security of knowing it is close at hand, then try to keep it out of view, or at least turn it upside down; keep it away from your direct reach, perhaps on the other side of the room. A recent study by Professor Bill Thornton and his colleagues at the University of Southern Maine demonstrated that when performing complex tasks that require our full attention even the mere presence of the experimenter's phone (not the participant's phone) led to distraction and worse performance.[15] In the same study, the presence of a student's silenced phone in a classroom had an equally negative impact on attention.

Finally, there are apps that will help you control your environment such as SelfControl, Freedom, KeepMeOut, Cold Turkey, FocalFilter, FocusMe, Training Wheels, LeechBlock, TinyFilter, Anti-Social, Freedom, and Stay-Focused. These apps block specified websites for a set amount of time or limit your daily use of stipulated websites. If you are constantly checking social media, programs such as Concentrate or Think can be set up to only allow you to open certain tools for certain tasks, thereby limiting potential distractors; FocusWriter, WriteRoom, or JDarkRoom block out all programs that are not directly related to writing reports or school papers. If you find yourself writing an email and then getting lost in your inbox, consider the app Compose, which lets you write and send an email without ever viewing your inbox.

↓ Boredom

Let's face it: working on critical assignments can be boring sometimes, especially when the alternative is so much more appealing—high-definition videos, immersive video games, and endless social media connections that are just a click away. One strategy to decrease boredom while working on an assignment is to spend some of your computer time standing rather than sitting. You can simply raise your computer by putting it on top of a box or you can even purchase a treadmill desk that allows you to stand or walk at your discretion. This has the added benefit that walking rather than sitting increases blood to the brain during challenging cognitive control tasks.[16] You can also listen to music, particularly songs you enjoy, as an easy way to increase your mood while focusing on a single task, as well as improve cognitive task performance.[17] Listening to familiar music also has been shown to reduce stress in medical and dental patients while at the same time increasing the efficiency of the medical professionals.[18] Of course, this has to be managed with the potentially distracting impact of external stimulation, even of music, on performance.[19]

As discussed in chapter 9, technology may be decreasing the time associated with the onset of boredom when single tasking as a result of our ever-escalating exposure to pervasive, high-frequency reward feedback; that is, the rapid pace of video games, texts, and email traffic is changing our tolerance for slower-paced activities. One strategy we propose to overcome this is to gradually increase time on task before allowing yourself to take a break. The idea is that by using time-delayed onset of breaks as rewards, you can take baby steps into building a greater tolerance for slower-paced reward cycles. You control the breaks, rather than the breaks controlling you.

When it comes to taking breaks, it is important to note that not all of them are created equal. Here are some thoughts on how we might use particularly effective breaks to stay on the primary task longer before boredom wins the battle. As University of Illinois psychology professors Atsunori Ariga and Alejandro Lleras explain: "Deactivating and reactivating your goals allows you to stay focused. From a practical standpoint, research suggests that, when faced with long tasks (such as studying before a final exam or doing your taxes), it is best to impose brief breaks. Brief mental breaks will actually help you stay focused on your task!"[20] So, even if breaks do not actually make your assignments less boring, the positive effects of combating fatigue and reducing stress will maintain focus, as the overall time engaged in your assignment becomes more rewarding.

Here are some ideas based on research studies for planning restorative, stress-reducing breaks, each of which will take you only a few minutes.

- Exercise—even for only twelve minutes—facilitates brain function and improves attention, as discussed in detail in chapter 10.[21]
- Train your eyes using the 20–20–20 rule: every twenty minutes take a twenty-second break and focus on objects twenty feet away. This changes your focal distance from inches to many feet and requires blood flow to brain areas that are not related to constant attention.[22]
- Expose yourself to nature. Consider using at least part of your break to get away from technology and spend a few minutes in a natural setting. Research has shown that just ten minutes in a natural environment can be restorative; even viewing pictures of nature can be restorative, as discussed in chapter 10.[23]
- Daydreaming, staring into space, doodling on paper, or any activity that takes you away from performing a specified task activates the "default mode network"—a network of interacting brain areas that most often indicate that you are daydreaming, thinking creatively, or just mind wandering—which is restorative for attention.[24]
- Short ten-minute naps have been shown to improve cognitive function. Longer naps work, too, as seen in a study of pilots who improved their reaction time after taking a thirty-minute nap.[25]
- Talking to other human beings, face to face or even on the telephone, reduces stress and has been shown to improve work performance.[26]
- Laugh! Read a joke book, look at comic strips, read a funny blog. A Loma Linda University study found that older adults who watched a funny video scored better on memory tests and showed reduced cortisol and increased endorphins and dopamine, meaning less stress and more energy and positive feelings.[27]
- Grab something to drink and a small snack.[28]
- Read a chapter in a fiction book. Recent research shows major brain shifts when reading immersive fiction.[29]

The bottom line is to pay attention to what you choose to do when you take a break at work, between classes at school, or have a few moments to put your feet up and relax at home (and this should really not always include your grabbing your smartphone, for all the reasons that we have discussed in this book!). Whatever relaxes you and takes you away from your overstimulating technological environment will help you reengage with greater arousal, more capacity for attention, and less susceptibility to being interrupted.

↓ Anxiety

In Part II, we discussed how technology has induced anxiety associated with FOMO, which then causes you to interrupt your work and reorient your attentional resources to the detriment of your performance on that all-important task. The strategies we learned in the previous scenario can also be applied here. Start by setting expectations; inform your colleagues that you have a new plan in which communications happen in pre-established intervals. There are many ways to do this, but here is what we recommend. First, send out an email (or text) to anyone that you connect with on a regular and continuing basis. Explain that you are working under a "90–20 plan" (or perhaps one of the plans we will discuss below) and that you will be off the grid for ninety minutes and then will check and answer all communications during your break time. If you regularly respond immediately on social media, post a notice there about your new work plan. Second, at the beginning of your first week of implementation, set up an automatic reply that spells out your plan. After a week most people will get it. Third, if you are working in an environment where other people also work (office, library, home) post a red sign that says "No Interruption Zone" and provide a time that you will be available. Make a green sign that says, "I Am Available Now to Be Interrupted." If necessary, you can also use apps to automatically reply to texts and phone calls and allow emergency calls through. This approach will diminish anxiety from FOMO by allowing you to not expect communication all of the time and to know that the critical ones will get through. Even if it is not effective in preventing your entire professional and social world from contacting you except when you want them to (the reality is that is not going to happen), it will start to diminish anxiety as you get used to a sparser, more controlled communication environment.

Here are two other options to consider that target anxiety in a more general manner. Consider engaging in meditation and mindfulness practices, as discussed in chapter 10, for its impact on the brain. A recent Johns Hopkins

University meta-analysis of forty-one randomized, controlled mindful meditation trials including 2,993 participants demonstrated moderate effect sizes for reducing anxiety.[30] Also, as previously discussed, engaging in a physical exercise is a great idea as meta-analyses of exercise programs with healthy adults found effects in reducing anxiety and increasing overall quality of life.[31] These results suggest the possibility that meditation and exercise may also diminish FOMO that drives many of us to constantly check in with technology, but this still needs to be validated with research.

STRATEGIES FOR UNINTERRUPTED SOCIALIZING

Look around any restaurant and you'll see phones on almost every table (accessibility). Most people are probably unaware that the presence of a phone, even if unanswered, negatively affects our relationships with others (poor metacognition). People are more likely to pause a face-to-face conversation to check an incoming text (anxiety) and look down at their phone for no reason at all (boredom) than ever before. Here are some ideas on how to make the most out of the time you have to socialize with your family and friends when they are in your presence rather than self-interrupting and degrading those important human connections.

↑ Metacognition

In chapter 6, we discussed how people bring devices into situations that used to be devoted to face-to-face connection, such as going on a date, attending a child's school function, going on vacation, and even having sex. In chapter 7, we explored the impact of technology on our relationships, including studies that suggested even the mere presence of any phone reduces closeness, connection, and conversation quality as well as reducing the extent to which individuals feel empathy and understanding from their partners, all of which negatively affects our relationships with others.[32] Mobile phones can reduce both our own empathy and our perception of other people's empathy toward us. A 2015 study of young adults found that "a substantial number of them have experienced mobile device use as disruptive to their face-to-face encounters."[33] It is clear from the advent of the "cellphone stack" that some of us do recognize the impact that a phone has on conversation.

In chapter 7, we discussed the concept of nomophobia—anxiety caused by not having your mobile phone on you at all times. Want to see if you have nomophobia? You can take a test that asks about when you first check your phone in the morning, how often you check it, where you keep it when you are sleeping, whether you take it with you to the bathroom (many people do!), and what you use it for.[34] It is likely that many of you will find you suffer from nomophobia. This knowledge about yourself may help motivate you to take better control of your smartphone use.

↓ Accessibility

The best strategy available to preserve the quality of "uninterrupted" interpersonal relationships is to simply have all people involved turn off their phones or remove them from the area of interaction. If that is too difficult, you might try using "tech breaks," allowing conversationalists to first check their phones and then turn them off, perhaps setting an alarm for fifteen minutes and allowing each person to check in for a minute, and then repeating the process. This can be an effective way of discouraging unplanned interruptions that lead to unsatisfying interpersonal interactions.

Another way to reduce accessibility is to create a "technology-free zone," which means no television, no smartphones, no devices of any kind. For example, Arianna Huffington has advocated making the bedroom a technology-free zone, imploring people to, "ban their tablets and smart phones from their bedrooms, for the sake of their health."[35] We heartily support technology-free zones as a way of avoiding a Distracted Mind and have talked to some parents who opt to make meal areas technology-free zones. As noted before, for some it is impossible to even consider having their smartphone out of direct sight, but if you couple the idea of technology-free times with technology-free zones, this strategy could work. For example, halfway through dinner a one-minute break could be declared where everyone gets to check in, which then could eventually be phased out as family members learn that they are not missing out if they don't check in during their dinnertime.

Not all technologies are equally disruptive to relationships. Personal devices may pull people away from live interaction but television and video games can be social experiences. Although the typical family no longer watches television together, if this is important to you, perhaps you could try to find one weekly show that you can view together as a family, and during that viewing you can declare the television location a personal-device-free zone, perhaps allowing personal tech breaks much as we use commercials for bathroom or snack breaks. We recommend that parents invoke two ways of encouraging unencumbered family interactions: weekly family meetings and weekly shared technology times that may include joint television watching or movie viewing or even better—interactive game playing. Family meetings should be about fifteen minutes (although they can be longer), and they should involve parents catching up on their children's activities during the week and sharing experiences in a positive way. What seems to work best is for everyone to sit down together at the table, on a couch or even on the floor to equalize height differences as best as possible, since taller people are seen as more dominant, and to silence and remove all personal devices from the area.[36]

↓ Boredom

In Part II, we shared research showing that many people reach for their phones in social situations that are boring or uninteresting. It is, we believe, important for us to relearn how not to be bored by situations that transpire more slowly, and to accept that not all conversations are going to be equally stimulating. And most importantly, we need to understand that disengaging immediately by hiding behind technology is a sure way to miss out on something interesting. If we let our attention be interrupted constantly by what is contained in our virtual world, we are going to miss some gems in our real world.

A good place to practice retraining ourselves is during meals. Numerous studies have examined the impact of family meals, and a recent meta-analysis found that frequent family meals are associated with more positive outcomes including better children and adolescent psychosocial health as well as improved family relationships.[37] With regard to boredom, you can help children or others enjoy regular mealtimes by making sure they are included in conversation but don't feel they are being "grilled."

↓ Anxiety

Many of the techniques that we recommended in earlier scenarios will be helpful in reducing anxiety in social situations that are caused by fear of missing out on another interaction that is going on elsewhere. As we've discussed, setting expectations is a good first step. If you make mealtimes technology free, as suggested earlier, let your friends and colleagues know that you won't be available during that time. If necessary, you can allow emergency calls using the techniques described earlier such as Essential Calls or Selective Silence, or the Do Not Disturb function coupled with the Allow Calls. Another option, so you aren't worried about missing an important text or call, is to use an app that auto-responds to texts and phone calls and informs the caller when you will be available.

STRATEGIES FOR GETTING A GOOD NIGHT'S SLEEP

Most of us know the value of a good night's sleep, but the pressures of life, school, and work push us to squeeze more out of our days by shortening our nights. We bring our laptops into the bedroom to get a little more work done or play a phone game in bed to amuse ourselves (boredom) before we fall asleep without realizing the impact it has on our all-important sleep patterns (poor metacognition). Then we leave our devices right next to our bed (accessibility), afraid to miss something during the night or not catch up on the latest activities first thing in the morning (anxiety). As we have shown earlier in this book, technology is particularly disruptive to our sleep, which is then harmful to our health and brain function.[38] Here are some strategies to help you regain control over a task that should get really get uninterrupted focus: going to sleep and staying asleep.

↑ Metacognition

Chapter 5 explained how even a single bad night's sleep can impair cognitive control and how ongoing sleep deprivation can have severe and long-term consequences. Chapter 6 showed how our technology is invading our bedrooms. Most young people (and many older ones, too) take their phones, tablets, and computers to bed with them and use them right up to the moment that they need to fall asleep, and nearly half of them awaken at night and check their phone, which they have left in plain sight by their bedside. Chapter 7 discussed how technology use prior to bedtime negatively affects sleep, particularly among children and adolescents. American adults have also experienced a decline in the number of hours of nightly sleep as well as in the quality of that sleep, with one factor being attributed to using their electronic devices within an hour of bedtime. Poor sleep is a major problem for children, adolescents, and adults, affecting their information processing, memory, and emotional states.

It is important that you are aware of the negative impact that technology has on your sleep. It is possible that unless you have someone monitoring your nighttime behavior you may not even be aware that you are not sleeping well. A recent *New York Times* article discussed the various options including several watch-based systems including Jawbone Up, Fitbit, Basis Peak, and Microsoft Band.[39] The article also described devices that are worn on the head or chest such as ResMedS+ that monitor many sleep activities. Finally, there are devices that you can attach to your bed such as Aura or Beddit, or you can go all the way and purchase a SleepIQ—which includes a sensing pad and a smartphone app to provide you with a nightly sleep score—an option when you purchase a Sleep Number bed.

↓ Accessibility

In terms of accessibility to technology and improving sleep, the ideal plan seems pretty straightforward: remove all technology from your bedroom. However, this will not be acceptable to many people, and so a more practical approach involves limiting accessibility, or even limiting the more disruptive aspects of nighttime technologies. The National Sleep Foundation strongly suggests that in order to get a good night's sleep you do the following for yourself and your children: (1) progressively reduce the light in the bedroom at least an hour or more before bedtime to allow maximal release of melatonin, (2) keep electronics out of the bedroom and turned off for at least an hour before bedtime, and (3) immerse yourself in as much daylight as you can during the daytime (meaning take breaks and go outside) to regulate your body's biochemistry.[40] For those who are not able to give up their devices entirely at night, a Mayo Clinic study suggests that you should dim your smartphone or tablet brightness settings and position that device at least fourteen inches from your face, which then reduces the amount of blue light entering the eyes and subsequently has minimal impact on melatonin release.[41] To avoid the blue light from your devices, consider software such as f.lux, which changes the color of your device screen light to match the time of day, providing warmer colors at night and allowing the melatonin to be released.[42]

A final comment about the intersection of technology and your sleep concerns your morning rituals. Earlier, we mentioned a study that found that eight in ten adult smartphone owners reach for their phone within fifteen minutes of awakening and six in ten check their phone as soon as they awaken. For young adults those figures are much higher.[43] If the phone is removed from the bedroom before going to sleep and placed in another location, this will limit accessibility and help you awaken slowly and gradually, which is far preferable to immediately checking your phone and putting your brain on high alert.

↓ Boredom

For many people, the smartphone or tablet serves to reduce boredom as one prepares to go to sleep. Lying down to go to sleep is not high on the list of exciting activities for many people, especially children. But you can make it less boring by enhancing your sleep environment. One option is to consider listening to music in the darkened room as you drift off to sleep. A recent meta-analysis of randomized controlled research has shown that for people with chronic sleep disorders—nightly interrupted sleep would qualify as a chronic sleep disorder—music improved sleep quality significantly.[44] Other research has shown similar results in aiding sleep quality among adults with sleep complaints.[45]

Although there is scant research on the impact of our "environment" on reducing the boredom involved in preparing to go to sleep, some interesting research suggests that altering your bedroom environment can make sleep more enjoyable. For example, there are beds that can be adjusted to your specific comfort level with separate controls for each side of the bed. In addition, some work suggests that sleep quality and attitude toward sleep can be improved by gradually lowering the room lights.[46]

↓ Anxiety

As described earlier, a study from Dr. Rosen's lab that found that strongest predictor of poor sleep quality was anxiety about missing out on an "important" electronic communication. This anxiety most likely led to the participants awakening at night to check for messages. This is a difficult issue and one that often rests on the definition of important communications. Once again, you will need to set expectations, keep people informed, and make sure you feel comfortable that you can be reached in an emergency. We would suggest that you specify a small set of people for whom a nighttime phone call is truly an emergency and then use an app such as Essential Calls or Selective Silence for Android devices, or by using the Do Not Disturb function coupled with the Allow Calls From option on iPhones, to determine which calls should be able to ring through even when the phone is on silent. You can also have your phone ring after multiple attempts and update your voicemail message so people know to call again if it is really urgent. Apps can auto-respond to a text messages stating that you can be reached via phone only for urgent messages.

SUMMARY

It should be clear that if you are having trouble keeping yourself from constantly being interrupted by technology, you are not alone: you are a typical modern-day high-tech user. We have provided you with numerous strategies to combat this negative influence on your Distracted Mind using four very common everyday situations: driving, completing important assignments, socializing, and sleeping. Each set of strategies focused on the four factors of the MVT model, helping you alleviate the stress on your cognitive control limitations: improving metacognition, decreasing the accessibility of interruptive technologies, decreasing your boredom, and minimizing your anxiety of missing out. There are undoubtedly other situations in which you find your attention being pulled away, and these same strategies should work to benefit you there as well. Regardless of the scenario, the suggested strategies are meant to be adapted to the specific situation, and by applying them you will get better at maintaining your focus on what is important rather than being interrupted by what is not critical at the moment.

Don't expect that changing your behavior will be easy. We have been susceptible to distractions and interruptions for our entire lives, but technology's impact on the Distracted Mind has caused us to overindulge. Our ability to control our Distracted Mind has spiraled out of control because of the technological game changers that we discussed in Part II. This is why some people will feel as though they have no choice but to attempt a digital detox. As we have said in this chapter, we do not feel that drastic solutions are the answer. The strategies we have presented are less extreme and promise more likelihood of success.

Changing your behavior may not be easy, but it is doable. If you find yourself being distracted, ask yourself the following questions:

- How might I increase my metacognitive view of how my own mind performs in a given situation, and in what ways are my actions not in line with how I should behave based on my goals and an understanding of my limitations?
- How might I change my physical environment to reduce accessibility of potential distractors?
- How might I assess whether I am self-interrupting because of boredom, and how might I make the task more interesting to stave off that boredom?
- How might I recognize when my actions are driven by anxiety about missing out on something in my virtual world, and what steps can I take to reduce that anxiety?

If you follow the strategies outlined in this chapter, you will go a long way to avoid the pitfalls of distractions and interruptions. This can be coupled with the approaches discussed in chapter 10 that are available now, or on the horizon, to improve our cognitive control, reduce goal interference, and aid our Distracted Mind.

Notes

CHAPTER 1: INTERFERENCE

1. W. C. Clapp and A. Gazzaley, "Distinct Mechanisms for the Impact of Distraction and Interruption on Working Memory in Aging," *Neurobiology of Aging* 33, no. 1 (2012): 134–148.

2. M. A. Killingsworth and D. T. Gilbert, "A Wandering Mind Is an Unhappy Mind," *Science*, 330, no. 6006 (2010): 932.

3. F. Coolidge and T. Wynn, "Executive Functions of the Frontal Lobes and the Evolutionary Ascendancy of Homo Sapiens," *Cambridge Archaeological Journal* 11, no. 2 (2001): 255–260; M. Tomasello and E. Herrmann, "Ape and Human Cognition: What's the Difference?" *Current Directions in Psychological Science* 19 (2010): 3–8.

4. S. Inoue and T. Matsuzawa, "Working Memory of Numerals in Chimpanzees," *Current Biology* 17, no. 23 (2007): R1004–R1005; N. Kawai and T. Matsuzawa, "Numerical Memory Span in a Chimpanzee," *Nature* 403, no. 6765 (2000): 39–40; M. J.-M. Macé, G. Richard, A. Delorme, and M. Fabre-Thorpe, "Rapid Categorization of Natural Scenes in Monkeys: Target Predictability and Processing Speed," *NeuroReport* 16, no. 4 (2005): 349–354; S. F. Sands and A. A. Wright, "Monkey and Human Pictorial Memory Scanning," *Science* 216, no. 4552 (1982): 1333–1334.

5. M. Anderson, *Technology Device Ownership: 2015*, Pew Research Center report, retrieved on March 2, 2016, from http://www.pewinternet.org/files/2015/10/PI_2015-10-29_device-ownership _FINAL.pdf; Pew Research Center, *U.S. Smartphone Use in 2015*, retrieved on March 2, 2016, from http://www.pewinternet.org/files/2015/03/PI_Smartphones_0401151.pdf.

6. "Global Mobile Statistics 2012, Part A: Mobile Subscribers, Handset Market Share, Mobile Operators," *mobithinking.com*, December 2012, http://mobithinking.com/mobile-marketing-tools/latest-mobile-stats.

7. L. M. Carrier, N. A. Cheever, L. D. Rosen, S. Benitez, and J. Chang, "Multitasking across Generations: Multitasking Choices and Difficulty Ratings in Three Generations of Americans," *Computers in Human Behavior* 25 (2009): 483–489.

8. J. Q. Anderson and L. Rainie, *Millennials Will Benefit and Suffer due to Their Hyperconnected Lives*, PEW Internet and American Life Project, 2012, http://pewinternet.org/~/media//Files/ Reports/2012/PIP_Future_of_Internet_2012_Young_brains_PDF.pdf; Carrier, Cheever, Rosen,

Benitez, and Chang, "Multitasking across Generations"; U. G. Foehr, *Media Multitasking among American Youth: Prevalence, Predictors, and Pairings: Report* (Menlo Park, CA: Kaiser Family Foundation, 2006), http://www.kff.org/entmedia/upload/7592.pdf; S. A. Brasel and J. Gips, "Media Multitasking Behavior: Concurrent Television and Computer Usage," *Cyberpsychology, Behavior, and Social Networking* 14, no. 9 (2011): 527–534; S. Kessler, "38% of College Students Can't Go 10 Minutes without Tech [STATS]," Mashable Tech, 2011, http://mashable.com/2011/05/31/college-tech-device-stats/.

9. T. Ahonen, "Main Trends in the Telecommunications Market," presentation at MoMo mobile conference, Kiev, Ukraine, http://www.citia.co.uk/content/files/50_44-887.pdf; "Anxiety UK Study Finds Technology Can Increase Anxiety," *AnxietyUK.org*, July 9, 2012, http://www.anxietyuk.org.uk/2012/07/for-some-with-anxiety-technology-can-increase-anxiety/; Lockout Mobile Security, "Mobile Mindset Study" (2012), https://www.mylookout.com/downloads/lookout-mobile-mindset-2012.pdf.

10. Harris Interactive, "Americans Work on Their Vacation: Half of Those Vacationing Will Work on Their Vacation, Including Checking Emails, Voicemails, and Taking Calls," July 28, 2011, http://www.harrisinteractive.com/NewsRoom/HarrisPolls/tabid/447/mid/1508/articleId/843/ctl/ReadCustom%20Default/Default.aspx.

11. Y. Hwang, H. Kim, and S. H. Jeong, "Why Do Media Users Multitask? Motives for General, Medium-Specific, and Content-Specific Types of Multitasking," *Computers in Human Behavior* 36 (2014): 542–548; S. Chinchanachokchai, B. R. Duff, and S. Sar, "The Effect of Multitasking on Time Perception, Enjoyment, and Ad Evaluation," *Computers in Human Behavior* 45 (2015): 185–191.

12. L. Yeykelis, J. J. Cummings, and B. Reeves, "Multitasking on a Single Device: Arousal and the Frequency, Anticipation, and Prediction of Switching between Media Content on a Computer," *Journal of Communication* 64, no. 1 (2014): 167–192.

13. B. C. Wittmann, N. Bunzeck, R. J. Dolan, and E. Düzel, "Anticipation of Novelty Recruits Reward System and Hippocampus While Promoting Recollection," *NeuroImage* 38, no. 1 (2007): 194–202.

14. O. Hikosaka, S. Yamamoto, M. Yasuda, and H. F. Kim, "Why Skill Matters," *Trends in Cognitive Sciences*, 17, no. 9 (2013): 434–441.

15. T. T. Hills, "Animal Foraging and the Evolution of Goal-Directed Cognition," *Cognitive Science* 30, no. 1 (2006): 3–41.

16. R. A. Wise, "Dopamine, Learning, and Motivation," *Nature Reviews Neuroscience* 5, no. 6 (2004): 483–494; M. van Schouwenburg, E. Aarts, and R. Cools, "Dopaminergic Modulation of Cognitive Control: Distinct Roles for the Prefrontal Cortex and the Basal Ganglia," *Current Pharmaceutical Design* 16, no. 18 (2010): 2026–2032; M. Wang, S. Vijayraghavan, and P. S. Goldman-Rakic, "Selective D2 Receptor Actions on the Functional Circuitry of Working Memory," *Science* 303 (2004): 853–856; M. Watanabe, T. Kodama, and K. Hikosaka, "Increase of Extracellular Dopamine in Primate Prefrontal Cortex During a Working Memory Task," *Journal of Neurophysiology* 78, no. 5 (1997): 2795–2798.

17. E. S. Bromberg-Martin and O. Hikosaka, "Midbrain Dopamine Neurons Signal Preference for Advance Information about Upcoming Rewards," *Neuron* 63, no. 1 (2009): 119–126.

18. T. T. Hills, "Animal Foraging."

19. P. Pirolli and S. Card, "Information Foraging," *Psychological Review* 106, no. 4 (1999): 643.

20. E. L. Charnov, "Optimal Foraging: The Marginal Value Theorem," *Theoretical Population Biology* 9, no. 2 (1976): 129–136.

21. M. H. Cassini, A. Kacelnik, and E. T. Segura, "The Tale of the Screaming Hairy Armadillo, the Guinea Pig, and the Marginal Value Theorem," *Animal Behavior* 39, no. 6 (1990: 1030–1050; R. J. Cowie, "Optimal Foraging in the Great Tits (*Parus Major*)," *Nature* 268 (1977): 137–139.

22. Pirolli and Card, "Information Foraging"; T. Hills, P. M. Todd, and R. L. Goldstone, "Priming and Conservation between Spatial and Cognitive Search," in *Proceedings of the 29th Annual Cognitive Science Society* (Austin: Cognitive Science Society, 2007), 359–364; P. E. Sandstrom, "An Optimal Foraging Approach to Information Seeking and Use," *Library Quarterly* (1994): 414–449; M. Dwairy, A. C. Dowell, and J. C. Stahl, "The Application of Foraging Theory to the Information Searching Behavior of General Practitioners," *BMC Family Practice*, 12, no. 1 (2011): 90.

CHAPTER 2: GOALS AND COGNITIVE CONTROL

1. R. Marois and J. Ivanoff, "Capacity Limits of Information Processing in the Brain," *Trends in Cognitive Sciences* 9, no. 6 (2005): 296–305.

2. J. M. Fuster, "Upper Processing Stages of the Perception-Action Cycle," *Trends in Cognitive Sciences* 8, no. 4 (2004): 143–145.

3. The term "perception-action cycle" was coined and popularized by Joaquin Fuster, but the concept was referred to by various other names going back to 1950. See J. M. Fuster, *The Prefrontal Cortex*, 2nd ed. (New York: Raven Press, 1989); J. M. Fuster, *Cortex and Mind: Unifying Cognition* (Oxford: Oxford University Press, 2003).

4. The patella reflex may be more accurately referred to as a sensation-action reflex, because the brain is not in the core loop.

5. F. L. Coolidge and T. Wynn, "Executive Functions of the Frontal Lobes and the Evolutionary Ascendancy of Homo Sapiens," *Cambridge Archeological Journal* 11, no. 2 (2001): 255–260.

6. N. J. Emery and N. S. Clayton, "The Mentality of Crows: Convergent Evolution of Intelligence in Corvids and Apes," *Science* 306, no. 5703 (2004): 1903–1907.

7. Quoted material in this paragraph is from Charles Sabine, "Senses Helped Animals Survive the Tsunami," NBC News with Brian Williams, http://www.nbcnews.com/id/6795562/ns/nbc _nightly_news_with_brian_williams/t/senses-helped-animals-survive-tsunami.

8. National Highway Traffic Safety Administration and Virginia Tech Transportation Institute, "Breakthrough Research on Real-World Driver Behavior Released," April 20, 2006, http://www .nhtsa.gov/Driving+Safety/Distracted+Driving+at+Distraction.gov/Breakthrough+Research +on+Real-World+Driver+Behavior+Released.

9. William James, *Principles of Psychology* (New York: Holt, 1890), 404.

10. S. J. Luck and S. P. Vecera, "Attention," in *Stevens' Handbook of Experimental Psychology*, vol. 1: *Sensation and Perception*, ed. H. Pasher and S. Yantis (New York: John Wiley, 2002), 235–286.

11. A. Baddeley, *Working Memory* (Oxford: Oxford University Press, 1986).

12. A. Gazzaley and A. C. Nobre, "Top-Down Modulation: Bridging Selective Attention and Working Memory," *Trends in Cognitive Sciences* 16, no. 2 (2012): 129–135.

13. D. Premack, "Human and Animal Cognition: Continuity and Discontinuity," *Proceedings of the National Academy of Sciences of the United States of America* 104, no. 35 (2007): 13861–13867.

CHAPTER 3: THE BRAIN AND CONTROL

1. A. Gazzaley and M. D'Esposito, "Unifying Prefrontal Cortex Function: Executive Control, Neural Networks, and Top-Down Modulation," in *The Human Frontal Lobes*, ed. B. Miller and J. Cummings (New York: Guilford, 2007).

2. G. Fritsch and E. Hitzig, "Uber die elektrische Erregbarkeit des Grosshiirns," *Archiv der Anatomie, Physiologie und Wissenschaftlichen Medizin* 37 (1870): 300–332.

3. B. J. Baars and N. M. Gage, *Cognition, Brain, and Consciousness: Introduction to Cognitive Neuroscience* (New York: Academic Press, 2010).

4. K. Semendeferi, H. Damasio, R. Frank, and G. W. Van Hoesen, "The Evolution of the Frontal Lobes: A Volumetric Analysis Based on Three-Dimensional Reconstructions of Magnetic Resonance Scans of Human and Ape Brains," *Journal of Human Evolution* 32, no. 4 (1997): 375–388.

5. H. J. Bigelow, "Dr. Harlow's Case of Recovery from the Passage of an Iron Bar through the Head," *American Journal of the Medical Sciences* 16, no. 39 (1850): 13–22. See also the Phineas Gage Information Page at http://www.uakron.edu/gage.

6. J. M. Harlow, "Recovery from the Passage of an Iron Bar through the Head," *Publications of the Massachusetts Medical Society Publications* 2 (1868): 327–346.

7. H. Damasio, T. Grabowski, R. Frank, A. M. Galaburda, and A. R. Damasio, "The Return of Phineas Gage: Clues about the Brain from the Skull of a Famous Patient," *Science* 264, no. 5162 (1994): 1102–1105.

8. Harlow, "Recovery from the Passage of an Iron Bar through the Head"; reprinted as J. M. Harlow, *Recovery from the Passage of an Iron Bar through the Head* (Boston: David Clapp & Son), 13.

9. G. A. Mashour, E. E. Walker, and R. L. Martuza, "Psychosurgery: Past, Present, and Future," *Brain Research Reviews* 48, no. 3 (2005): 409–419.

10. W. Freeman and J. W. Watts, "Physiological Psychology," *Annual Review of Physiology* 6, no. 1 (1944): 517–542.

11. Freeman and Watts, "Physiological Psychology."

12. A. L. Benton, "Differential Behavioral Effects in Frontal Lobe Disease," *Neuropsychologia* 6, no. 1 (1968): 53–60; A. R. Luria, *Human Brain and Psychological Processes* (New York: Harper & Row, 1968); B. Milner, "Effects of Different Brain Regions on Card Sorting," *Archives of Neurology* 9 (1963): 90–100.

13. P. T. Schoenemann, M. J. Sheehan, and L. D. Glotzer, "Prefrontal White Matter Volume Is Disproportionately Larger in Humans Than in Other Primates," *Nature Neuroscience* 8, no. 2 (2005): 242–252.

14. A. Gazzaley and M. D'Esposito, "Neural Networks: An Empirical Neuroscience Approach toward Understanding Cognition," *Cortex* 42, no. 7 (2006): 1037–1040.

15. J. van Whye, "The History of Phrenology on the Web" (2004), http://www.historyofphrenology.org.uk/.

16. E. A. Berker, A. H. Berker, and A. Smith, "Translation of Broca's 1865 Report: Localization of Speech in the Third Left Frontal Convolution," *Archives of Neurology* 43, no. 10 (1986): 1065–1072.

17. R. M. Sabbatini, "Phrenology: The History of Brain Localization," *Brain and Mind* 1 (1997), http://www.cerebromente.org.br/n01/frenolog/frenologia.htm.

18. B. Tizard, "Theories of Brain Localization from Flourens to Lashley," *Medical History* 3 (1959): 132–145.

19. J. M. Fuster, *Cortex and Mind: Unifying Cognition* (Oxford: Oxford University Press, 2003).

20. M. Mesulam, "A Cortical Network for Directed Attention and Unilateral Neglect," *Annals of Neurology* 10, no. 4 (1981): 309–325.

21. Gazzaley and D'Esposito, "Unifying Prefrontal Cortex Function."

22. A. Gazzaley, J. W. Cooney, K. McEvoy, R. T. Knight, and M. D'Esposito, "Top-Down Enhancement and Suppression of the Magnitude and Speed of Neural Activity," *Journal of Cognitive Neuroscience* 17, no. 3 (2005): 507–517.

23. E. K. Miller and J. D. Cohen, "An Integrative Theory of Prefrontal Cortex Function," *Annual Review of Neuroscience* 24, no. 1 (2001): 167–202.

24. The left brain represents the right world, so the right visual cortex corresponds to the left visual field, where our ancestor's attention is focused.

25. Gazzaley et al., "Top-Down Enhancement."

26. J. Z. Chadick, T. P. Zanto, and A. Gazzaley, "Structural and Functional Differences in Medial Prefrontal Cortex Underlie Distractibility and Suppression Deficits in Ageing," *Nature Communications* 5 (2014): 4223.

27. J. Rissman, A. Gazzaley, and M. D'Esposito, "Measuring Functional Connectivity during Distinct Stages of a Cognitive Task," *Neuroimage* 23, no. 2 (2004): 752–763.

28. A. Gazzaley, J. Rissman, and M. D'Esposito, "Functional Connectivity during Working Memory Maintenance," *Cognitive, Affective, and Behavioral Neuroscience* 4, no. 4 (2004): 580–599.

29. T. P. Zanto, M. T. Rubens, A. Thangavel, and A. Gazzaley, "Causal Role of the Prefrontal Cortex in Top-Down Modulation of Visual Processing and Working Memory," *Nature Neuroscience* 14, no. 5 (2011): 656–661.

30. C. F. Jacobsen, "Studies of Cerebral Function in Primates," *Comparative Psychology Monographs* 13 (1938): 1–68.

31. J. M. Fuster and G. E. Alexander, "Neuron Activity Related to Short-Term Memory," *Science* 173, no. 3997 (1971): 652–654; K. Kubota and H. Niki, "Prefrontal Cortical Unit Activity and Delayed Alternation Performance in Monkeys," *Journal of Neurophysiology* 34 (1971): 337–347.

32. P. S. Goldman-Rakic, "Cellular Basis of Working Memory," *Neuron* 14, no. 3 (1995): 477–485.

33. J. M. Fuster, R. H. Bauer, and J. P. Jervey, "Functional Interactions between Inferotemporal and Prefrontal Cortex in a Cognitive Task," *Brain Research* 330, no. 2 (1985): 299–307.

34. W. C. Clapp, M. T. Rubens, and A. Gazzaley, "Mechanisms of Working Memory Disruption by External Interference," *Cerebral Cortex* 20, no. 4 (2009): 859–872.

35. P. E. Dux, J. Ivanoff, C. L. Asplund, and R. Marois, "Isolation of a Central Bottleneck of Information Processing with Time-Resolved fMRI," *Neuron* 52, no. 6 (2006): 1109–1120.

36. R. Kanai, M. Y. Dong, B. Bahrami, and G. Rees, "Distractibility in Daily Life Is Reflected in the Structure and Function of Human Parietal Cortex," *Journal of Neuroscience* 31, no. 18 (2011): 6620–6626.

CHAPTER 4: CONTROL LIMITATIONS

1. R. Desimone and J. Duncan, "Neural Mechanisms of Selective Visual Attention," *Annual Review of Neuroscience* 18, no. 1 (1995): 193–222.

2. T. P. Zanto and A. Gazzaley, "Neural Suppression of Irrelevant Information Underlies Optimal Working Memory Performance," *Journal of Neuroscience* 29, no. 10 (2009): 3059–3066.

3. E. K. Vogel, A. W. McCollough, and M. G. Machizawa, "Neural Measures Reveal Individual Differences in Controlling Access to Working Memory," *Nature* 438, no. 7067 (2005): 500–503.

4. A. M. Glenberg, J. L. Schroeder, and D. A. Robertson, "Averting the Gaze Disengages the Environment and Facilitates Remembering," *Memory and Cognition* 26 (1998): 651–658.

5. P. E. Wais, M. T. Rubens, J. Boccanfuso, and A. Gazzaley, "Neural Mechanisms Underlying the Impact of Visual Distraction on Retrieval of Long-Term Memory," *Journal of Neuroscience* 30, no. 25 (2010): 8541–8550.

6. P. E. Wais, O. Y. Kim, and A. Gazzaley, "Distractibility during Episodic Retrieval Is Exacerbated by Perturbation of Left Ventrolateral Prefrontal Cortex," *Cerebral Cortex* 22, no. 3 (2011): 717–724.

7. P. E. Wais and A. Gazzaley, "The Impact of Auditory Distraction on Retrieval of Visual Memories," *Psychonomic Bulletin and Review* 1, no. 6 (2011): 1090–1097.

8. M. A. Killingsworth and D. T. Gilbert, "A Wandering Mind Is an Unhappy Mind," *Science* 330, no. 6006 (2010): 932–932.

9. M. D. Mrazek, J. Smallwood, M. S. Franklin, B. Baird, J. M. Chin, and J. W. Schooler, "The Role of Mind-Wandering in Measurements of General Aptitude," *Journal of Experimental Psychology: General* 141 (2012): 788–798.

10. C. E. Rolle, B. Voytek, and A. Gazzaley, "Examining the Performance of the iPad and Xbox Kinect for Cognitive Science Research," *Games for Health Journal* 4, no. 3 (2015): 221–224.

11. B. S. Oken, M. C. Salinsky, and S. M. Elsas, "Vigilance, Alertness, or Sustained Attention: Physiological Basis and Measurement," *Clinical Neurophysiology* 117, no. 9 (2006): 1885–1901.

12. S. Bioulac, S. Lallemand, C. Fabrigoule, A. L. Thoumy, P. Philip, and M. P. Bouvard, "Video Game Performances Are Preserved in ADHD Children Compared with Controls," *Journal of Attention Disorders* 18, no. 6 (2014): 542–550.

13. K. L. Shapiro, J. E. Raymond, and K. M. Arnell, "The Attentional Blink," *Trends in Cognitive Sciences*, 1, no. 8 (1997): 291–296.

14. K. Fukuda and E. K. Vogel, "Human Variation in Overriding Attentional Capture," *Journal of Neuroscience* 29, no. 27 (2009): 8726–8733; K. Fukuda and E. K. Vogel, "Individual Differences in Recovery Time from Attentional Capture," *Psychological Science* 22, no. 3 (2011): 361–368.

15. T. F. Brady, T. Konkle, and G. A. Alvarez, "A Review of Visual Memory Capacity: Beyond Individual Items and toward Structured Representations," *Journal of Vision* 11, no. 5 (2011): 4.

16. G. A. Miller, "The Magical Number Seven, Plus or Minus Two: Some Limits on Our Capacity for Processing Information," *Psychological Review* 63, no. 2 (1956): 81.

17. N. Cowan, "The Magical Number 4 in Short-Term Memory: A Reconsideration of Mental Storage Capacity," *Behavioral and Brain Sciences* 24 (2001): 87–185.

18. S. J. Luck and E. K. Vogel, "The Capacity of Visual Working Memory for Features and Conjunctions," *Nature* 390, no. 6657 (1997): 279–281.

19. M. Daneman and P. A. Carpenter, "Individual Differences in Working Memory and Reading," *Journal of Verbal Learning and Verbal Behavior* 19, no. 4 (1980): 450–466; A. R. Conway, M. J. Kane, and R. W. Engle, "Working Memory Capacity and Its Relation to General Intelligence," *Trends in Cognitive Sciences* 7, no. 12 (2003): 547–552.

20. T. F. Brady, T. Konkle, J. Gill, A. Oliva, and G. A. Alvarez, "Visual Long-Term Memory Has the Same Limit on Fidelity as Visual Working Memory," *Psychological Science* 24, no. 6 (2013): 981–990.

21. W. C. Clapp, M. T. Rubens, and A. Gazzaley, "Mechanisms of Working Memory Disruption by External Interference," *Cerebral Cortex* 20, no. 4 (2009): 859–872.

22. W. C. Clapp and A. Gazzaley, "Distinct Mechanisms for the Impact of Distraction and Interruption on Working Memory in Aging," *Neurobiology of Aging* 33, no. 1 (2012): 134–148.

23. Vogel, McCollough, and Machizawa, "Neural Measures Reveal Individual Differences in Controlling Access to Working Memory."

24. Bernard I. Witt and Ward Lambert, IBM Operating System/360 Concepts and Facilities, http://bitsavers.informatik.uni-stuttgart.de/pdf/ibm/360/os/R01-08/C28-6535-0_OS360_Concepts _and_Facilities_1965.pdf.

25. Jesus Diaz, "How Multitasking Works in the New iPhone OS 4.0," *Gizmodo.com*, April 8, 2010, http://gizmodo.com/5512656/how-multitasking-works-in-the-new-iphone-os-40.

26. Diaz, "How Multitasking Works."

27. Clapp, Rubens, and Gazzaley, "Mechanisms of Working Memory Disruption by External Interference"; A. S. Berry, T. P. Zanto, A. M. Rutman, W. C. Clapp, and A. Gazzaley, "Practice-Related Improvement in Working Memory Is Modulated by Changes in Processing External Interference," *Journal of Neurophysiology* 102, no. 3 (2009): 1779–1789.

28. J. A. Anguera, J. Boccanfuso, J. L. Rintoul, O. Al-Hashimi, F. Faraji, J. Janowich, E. Kong, Y. Larraburo, C. Rolle, E. Johnston, and A. Gazzaley, "Video Game Training Enhances Cognitive Control in Older Adults," *Nature* 501, no. 7465 (2013): 97–101.

29. J. S. Rubinstein, D. E. Meyer, and J. E. Evans, "Executive Control of Cognitive Processes in Task Switching," *Journal of Experimental Psychology: Human Perception and Performance* 27, no. 4 (2001): 763.

CHAPTER 5: VARIATIONS AND FLUCTUATIONS

1. J. R. Best, P. H. Miller, and L. L. Jones, "Executive Functions after Age 5: Changes and Correlates," *Developmental Review* 29, no. 3 (2009): 180–200; M. C. Davidson, D. Amso, L. C. Anderson, and A. Diamond, "Development of Cognitive Control and Executive Functions from 4 to 13 Years: Evidence from Manipulations of Memory, Inhibition, and Task Switching," *Neuropsychologia* 44, no. 11 (2006): 2037–2078.

2. S. A. Bunge and S. B. Wright, "Neurodevelopmental Changes in Working Memory and Cognitive Control," *Current Opinion in Neurobiology* 17, no. 2 (2007): 243–250.

3. J. N. Giedd, J. Blumenthal, N. O. Jeffries, F. X. Castellanos, H. Liu, A. Zijdenbos, et al., "Brain Development during Childhood and Adolescence: A Longitudinal MRI Study," *Nature Neuroscience* 2, no. 10 (1999): 861–863; N. Gogtay, J. N. Giedd, L. Lusk, K. M. Hayashi, D. Greenstein, A. C. Vaituzis, T. F. Nugent, et al., "Dynamic Mapping of Human Cortical Development during Childhood through Early Adulthood," *Proceedings of the National Academy of Sciences of the United States of America* 101, no. 21 (2004): 8174–8179.

4. D. M. Lane and D. A. Pearson, "The Development of Selective Attention," *Merrill-Palmer Quarterly* (1982): 317–337; L. M. Trick and J. T. Enns, "Lifespan Changes in Attention: The Visual Search Task," *Cognitive Development* 13, no. 3 (1998): 369–386; J. T. Enns and S. Cameron, "Selective Attention in Young Children: The Relations between Visual Search, Filtering, and Priming," *Journal of Experimental Child Psychology* 44, no. 1 (1987): 38–63.

5. J. T. Enns and J. S. Girgus, "Developmental Changes in Selective and Integrative Visual Attention," *Journal of Experimental Child Psychology* 40, no. 2 (1985): 319–337.

6. L. P. McAvinue, T. Habekost, K. A. Johnson, S. Kyllingsbæk, S. Vangkilde, C. Bundesen, and I. H. Robertson, "Sustained Attention, Attentional Selectivity, and Attentional Capacity across the Lifespan," *Attention, Perception, and Psychophysics* 74, no. 8 (2012): 1570–1582.

7. See, e.g., http://www.playmaker.org.

8. S. A. Bunge and S. B. Wright, "Neurodevelopmental Changes in Working Memory and Cognitive Control," *Current Opinion in Neurobiology* 17, no. 2 (2007): 243–250.

9. F. R. Manis, D. P. Keating, and F. J. Morrison, "Developmental Differences in the Allocation of Processing Capacity," *Journal of Experimental Child Psychology* 29, no. 1 (1980): 156–169; C. Karatekin, "Development of Attentional Allocation in the Dual Task Paradigm," *International Journal of Psychophysiology* 52, no. 1 (2004): 7–21.

10. T. Zanto and A. Gazzaley, "Aging and Attention," in *Handbook of Attention*, ed. A. C. Nobre and S. Kastner (Oxford: Oxford University Press, 2014), 927–971.

11. N. Raz, "Aging of the Brain and Its Impact on Cognitive Performance: Integration of Structural and Functional Findings," in *The Handbook of Aging and Cognition*, 2nd ed., ed. F. I. M. Craik and T. A. Salthouse (New York: Erlbaum, 2000), 1–90.

12. Zanto and Gazzaley, "Aging and Attention."

13. L. Hasher, R. T. Zacks, and C. P. May, "Inhibitory Control, Circadian Arousal, and Age," in *Attention and Performance*, vol. 17 (Cambridge, MA: MIT Press, 1999), 653–675.

14. A. Gazzaley, J. W. Cooney, J. Rissman, and M. D'Esposito, "Top-Down Suppression Deficit Underlies Working Memory Impairment in Normal Aging," *Nature Neuroscience* 8, no. 10 (2005): 1298–1300.

15. T. P. Zanto, B. Toy, and A. Gazzaley, "Delays in Neural Processing during Working Memory Encoding in Normal Aging," *Neuropsychologia* 48, no. 1 (2010): 13–25; T. P. Zanto, P. Pan, H. Liu, J. Bollinger, A. C. Nobre, and A. Gazzaley, "Age-Related Changes in Orienting Attention in Time," *Journal of Neuroscience* 31, no. 35 (2011): 12461–12470; W. C. Clapp and A. Gazzaley, "Distinct Mechanisms for the Impact of Distraction and Interruption on Working Memory in Aging," *Neurobiology of Aging* 33, no. 1 (2012): 134–148.

16. J. Z. Chadick, T. P. Zanto, and A. Gazzaley, "Structural and Functional Differences in Medial Prefrontal Cortex Underlie Distractibility and Suppression Deficits in Aging," *Nature Communications* 5 (2014): 4223.

17. A. Gazzaley, W. Clapp, J. Kelley, K. McEvoy, R. T. Knight, and M. D'Esposito, "Age-Related Top-Down Suppression Deficit in the Early Stages of Cortical Visual Memory Processing," *Proceedings of the National Academy of Sciences* 105, no. 35 (2008): 13122–13126.

18. Gazzaley et al., "Age-Related Top-Down Suppression Deficit"; P. E. Wais, M. T. Rubens, J. Boccanfuso, and A. Gazzaley, "Neural Mechanisms Underlying the Impact of Visual Distraction on Retrieval of Long-Term Memory," *Journal of Neuroscience* 30, no. 25 (2010): 8541–8550.

19. Clapp and Gazzaley, "Distinct Mechanisms for the Impact of Distraction and Interruption"; J. Mishra, T. Zanto, A. Nilakantan, and A. Gazzaley, "Comparable Mechanisms of Working Memory Interference by Auditory and Visual Motion in Youth and Aging," *Neuropsychologia* 51, no. 10 (2013): 1896–1906.

20. T. S. Braver and R. West, "Working Memory, Executive Control, and Aging," *Handbook of Aging and Cognition* 3 (2008): 311–372; L. Hasher, C. Chung, C. P. May, and N. Foong, "Age, Time of Testing, and Proactive Interference," *Canadian Journal of Experimental Psychology* 56, no. 3 (2002): 200.

21. F. I. Craik and E. Dirkx, "Age-Related Differences in Three Tests of Visual Imagery," *Psychology and Aging* 7, no. 4 (1992): 661.

22. J. Kalkstein, K. Checksfield, J. Bollinger, and A. Gazzaley, "Diminished Top-Down Control Underlies a Visual Imagery Deficit in Normal Aging," *Journal of Neuroscience* 31, no. 44 (2011): 15768–15774.

23. Zanto and Gazzaley, "Aging and Attention."

24. Clapp and Gazzaley, "Distinct Mechanisms for the Impact of Distraction and Interruption"; Gazzaley et al., "Age-Related Top-Down Suppression Deficit."

25. Y. Gazes, B. C. Rakitin, C. Habeck, J. Steffener, and Y. Stern, "Age Differences of Multivariate Network Expressions during Task-Switching and Their Associations with Behavior," *Neuropsychologia* 50, no. 14 (2012): 3509–3518; D. J. Madden, M. C. Costello, N. A. Dennis, S. W. Davis, A. M. Shepler, J. Spaniol, et al., "Adult Age Differences in Functional Connectivity during Executive Control," *Neuroimage* 52, no. 2 (2010): 643–657; B. T. Gold, D. K. Powell, L. Xuan, G. A. Jicha, and C. D. Smith, "Age-Related Slowing of Task Switching Is Associated with Decreased Integrity of Frontoparietal White Matter," *Neurobiology of Aging* 31, no. 3 (2010): 512–522.

26. R. Helman, M. Greenwald, C. Copeland, and J. VanDerhei, "The 2010 Retirement Confidence Survey: Confidence Stabilizing, but Preparations Continue to Erode," *EBRI Issue Brief 340* (Washington, DC: Employee Benefit Research Institute, 2010).

27. A. Wingfield and E. A. L. Stine-Morrow, "Language and Speech," in *The Handbook of Aging and Cognition*, 359–416.

28. For these particular metrics, peak performance is indicated by the lowest point.

29. Unpublished data from Gazzaley Lab.

30. J. A. Anguera, J. Boccanfuso, J. L. Rintoul, O. Al-Hashimi, F. Faraji, J. Janowich, et al., "Video Game Training Enhances Cognitive Control in Older Adults," *Nature* 501, no. 7465 (2013): 97–101.

31. W. F. Chaplin, O. P. John, and L. R. Goldberg, "Conceptions of States and Traits: Dimensional Attributes with Ideals as Prototypes," *Journal of Personality and Social Psychology* 54, no. 4 (1988): 541.

32. D. I. Boomsma, "Genetic Analysis of Cognitive Failures (CFQ): A Study of Dutch Adolescent Twin Pairs and Their Parents," *European Journal of Personality* 12 (1998): 321–330.

33. R. Ratcliff and H. P. Van Dongen, "Sleep Deprivation Affects Multiple Distinct Cognitive Processes," *Psychonomic Bulletin and Review* 16, no. 4 (2009): 742–751.

34. J. S. Durmer and D. F. Dinges, "Neurocognitive Consequences of Sleep Deprivation," *Seminars in Neurology* 25, no. 1 (2005): 117–129.

35. C. E. Sexton, A. B. Storsve, K. B. Walhovd, H. Johansen-Berg, and A. M. Fjell, "Poor Sleep Quality Is Associated with Increased Cortical Atrophy in Community-Dwelling Adults," *Neurology* 83, no. 11 (2014): 967–973.

36. K. M. Edwards, R. Kamat, L. M. Tomfohr, S. Ancoli-Israel, and J. E. Dimsdale, "Obstructive Sleep Apnea and Neurocognitive Performance: The Role of Cortisol," *Sleep Medicine* 15, no. 1 (2014): 27–32; E. Gaio, P. DeYoung, A. R. Elliott, T. Limberg, A. Villa, M. Baylon, R. Sison-Tojino, et al., "High Prevalence of Obstructive Sleep Apnea in Patients with Moderate to Severe COPD," *American Journal of Respiratory Critical Care Medicine* 189 (2014): A5844–A5844.

37. E. H. Telzer, A. J. Fuligni, M. D. Lieberman, and A. Galván, "The Effects of Poor Quality Sleep on Brain Function and Risk Taking in Adolescence," *Neuroimage* 71 (2013): 275–283.

38. R. Gruber, J. Cassoff, S. Frenette, S. Wiebe, and J. Carrier, "Impact of Sleep Extension and Restriction on Children's Emotional Lability and Impulsivity," *Pediatrics* 130, no. 5 (2012): e1155–e1161.

39. L. Wade, "Kids with More Sleep Cope Better," *CNN.com*, October 16, 2012, http://www.cnn.com/2012/10/15/health/kids-sleep/index.html.

40. R. M. Yerkes and J. D. Dodson, "The Relation of Strength of Stimulus to Rapidity of Habit-Formation," *Journal of Comparative Neurology and Psychology* 18, no. 5 (1908): 459–482.

41. M. A. Staal, "Stress, Cognition, and Human Performance: A Literature Review and Conceptual Framework," *NASA Technical Memorandum*, 212824 (2004): 1–86.

42. M. J. Dry, N. R. Burns, T. Nettelbeck, A. L. Farquharson, and J. M. White, "Dose-Related Effects of Alcohol on Cognitive Functioning," *PLoS ONE* 7, no. 11 (2012): e50977; T. L. Martin, P. A. Solbeck, D. J. Mayers, R. M. Langille, Y. Buczek, and M. R. Pelletier, "A Review of Alcohol-Impaired Driving: The Role of Blood Alcohol Concentration and Complexity of the Driving Task," *Journal of Forensic Sciences* 58, no. 5 (2013): 1238–1250.

43. D. A. Brodeur and M. Pond, "The Development of Selective Attention in Children with Attention Deficit Hyperactivity Disorder," *Journal of Abnormal Child Psychology* 29, no. 3 (2001): 229–239; S. Forster, D. J. Robertson, A. Jennings, P. Asherson, and N. Lavie, "Plugging the Attention Deficit: Perceptual Load Counters Increased Distraction in ADHD," *Neuropsychology* 28, no. 1 (2014): 91–97.

44. C. S. Carter, P. Krener, M. Chaderjian, C. Northcutt, and V. Wolfe, "Abnormal Processing of Irrelevant Information in Attention Deficit Hyperactivity Disorder," *Psychiatry Research* 56, no. 1 (1995): 59–70; B. M. Ben-David, L. L. Nguyen, and P. H. van Lieshout, "Stroop Effects in Persons with Traumatic Brain Injury: Selective Attention, Speed of Processing, or Color-Naming? A

Meta-Analysis," *Journal of the International Neuropsychological Society* 17, no. 2 (2011): 354–363; A. M. Epp, K. S. Dobson, D. J. Dozois, and P. A. Frewen, "A Systematic Meta-Analysis of the Stroop Task in Depression," *Clinical Psychology Review* 32, no. 4 (2012): 316–328; J. M. Cisler, K. B. Wolitzky-Taylor, T. G. Adams, K. A. Babson, C. L. Badour, and J. L. Willems, "The Emotional Stroop Task and Posttraumatic Stress Disorder: A Meta-Analysis," *Clinical Psychology Review* 31, no. 5 (2011): 817–828; A. R. Moradi, M. R. Taghavi, H. T. Neshat Doost, W. Yule, and T. Dalgleish, "Performance of Children and Adolescents with PTSD on the Stroop Color-Naming Task," *Psychological Medicine* 29, no. 2 (1999): 415–419; A. Henik and R. Salo, "Schizophrenia and the Stroop Effect," *Behavioral and Cognitive Neuroscience Reviews* 3, no. 1 (2004): 42–59; B. M. Ben-David, A. Tewari, V. Shakuf, and P. H. van Lieshout, "Stroop Effects in Alzheimer's Disease: Selective Attention Speed of Processing, or Color-Naming? A Meta-Analysis," *Journal of Alzheimer's Disease* 38, no. 4 (2014): 923–938.

45. A. Christakou, C. M. Murphy, K. Chantiluke, A. I. Cubillo, A. B. Smith, V. Giampietro, et al., "Disorder-Specific Functional Abnormalities during Sustained Attention in Youth with Attention Deficit Hyperactivity Disorder (ADHD) and with Autism," *Molecular Psychiatry* 18, no. 2 (2013): 236–244.

46. M. A. Jenkins, P. J. Langlais, D. Delis, and R. A. Cohen, "Attentional Dysfunction Associated with Posttraumatic Stress Disorder among Rape Survivors," *Clinical Neuropsychologist* 14, no. 1 (2000): 7–12; J. J. Vasterling, K. Brailey, J. I. Constans, and P. B. Sutker, "Attention and Memory Dysfunction in Posttraumatic Stress Disorder," *Neuropsychology* 12, no. 1 (1998): 125; M.-L. Meewisse, M. J. Nijdam, G.-J. de Vries, B. P. R. Gersons, R. J. Kleber, P. G. van der Velden, A.-J. Roskam, et al., "Disaster-Related Posttraumatic Stress Symptoms and Sustained Attention: Evaluation of Depressive Symptomatology and Sleep Disturbances as Mediators," *Journal of Traumatic Stress* 18, no. 4 (2005): 299–302.

47. J. Whyte, M. Polansky, M. Fleming, H. B. Coslett, and C. Cavallucci, "Sustained Arousal and Attention after Traumatic Brain Injury," *Neuropsychologia* 33, no. 7 (1995): 797–813; P. P. Roy-Byrne, H. Weingartner, L. M. Bierer, K. Thompson, and R. M. Post, "Effortful and Automatic Cognitive Processes in Depression," *Archives of General Psychiatry* 43, no. 3 (1986): 265–267.

48. R. J. Perry and J. R. Hodges, "Attention and Executive Deficits in Alzheimer's Disease: A Critical Review," *Brain* 122, no. 3 (1999): 383–404.

49. E. G. Willcutt, A. E. Doyle, J. T. Nigg, S. V. Faraone, and B. F. Pennington, "Validity of the Executive Function Theory of Attention-Deficit/Hyperactivity Disorder: A Meta-Analytic Review," *Biological Psychiatry* 57, no. 11 (2005): 1336–1346.

50. M. D. Veltmeyer, C. R. Clark, A. C. McFarlane, K. A. Moores, R. A. Bryant, and E. Gordon, "Working Memory Function in Post-traumatic Stress Disorder: An Event-Related Potential Study," *Clinical Neurophysiology* 120, no. 6 (2009): 1096–1106; M. E. Shaw, K. A. Moores, R. C. Clark, A. C. McFarlane, S. C. Strother, R. A. Bryant, et al., "Functional Connectivity Reveals Inefficient Working Memory Systems in Post-traumatic Stress Disorder," *Psychiatry Research: Neuroimaging* 172, no. 3 (2009): 235–241.

51. T. W. McAllister, L. A. Flashman, M. B. Sparling, and A. J. Saykin, "Working Memory Deficits after Traumatic Brain Injury: Catecholaminergic Mechanisms and Prospects for Treatment—A Review," *Brain Injury* 18, no. 4 (2004): 331–350.

52. C. Christodoulou, J. DeLuca, J. H. Ricker, N. K. Madigan, B. M. Bly, G. Lange, et al., "Functional Magnetic Resonance Imaging of Working Memory Impairment after Traumatic Brain Injury," *Journal of Neurology, Neurosurgery, and Psychiatry* 71, no. 2 (2001): 161–168.

53. L. Pelosi, T. Slade, L. D. Blumhardt, and V. K. Sharma, "Working Memory Dysfunction in Major Depression: An Event-Related Potential Study," *Clinical Neurophysiology* 111, no. 9 (2000): 1531–1543; E. J. Rose and K. P. Ebmeier, "Pattern of Impaired Working Memory during Major Depression," *Journal of Affective Disorders* 90, no. 2 (2006): 149–161;S. Belleville, I. Peretz, and D. Malenfant, "Examination of the Working Memory Components in Normal Aging and in Dementia of the Alzheimer Type," *Neuropsychologia* 34, no. 3 (1996): 195–207.

54. N. Honzel, T. Justus, and D. Swick, "Posttraumatic Stress Disorder Is Associated with Limited Executive Resources in a Working Memory Task," *Cognitive, Affective, and Behavioral Neuroscience* 14, no. 2 (2014): 792–804; R. L. Aupperle, A. J. Melrose, M. B. Stein, and M. P. Paulus, "Executive Function and PTSD: Disengaging from Trauma," *Neuropharmacology* 62, no. 2 (2012): 686–694; I. A. Rasmussen Jr., J. Xu, I. K. Antonsen, J. Brunner, T. Skandsen, D. E. Axelson, et al., "Simple Dual Tasking Recruits Prefrontal Cortices in Chronic Severe Traumatic Brain Injury Patients, but not in Controls," *Journal of Neurotrauma* 25, no. 9 (2008): 1057–1070; G. S. Alexopoulos, D. N. Kiosses, S. Klimstra, B. Kalayam, and M. L. Bruce, "Clinical Presentation of the 'Depression–Executive Dysfunction Syndrome' of Late Life," *American Journal of Geriatric Psychiatry* 10, no. 1 (2002): 98–106; E. G. Willcutt, A. E. Doyle, J. T. Nigg, S. V. Faraone, and B. F. Pennington, "Validity of the Executive Function Theory of Attention-Deficit/ Hyperactivity Disorder: A Meta-Analytic Review," *Biological Psychiatry* 57, no. 11 (2005): 1336–1346; J. A. Foley, R. Kaschel, R. H. Logie, and S. Della Sala, "Dual-Task Performance in Alzheimer's Disease, Mild Cognitive Impairment, and Normal Aging," *Archives of Clinical Neuropsychology* 39, no. 1 (2011): 23–31.

55. A. F. Pettersson, E. Olsson, and L. O. Wahlund, "Effect of Divided Attention on Gait in Subjects With and Without Cognitive Impairment," *Journal of Geriatric Psychiatry and Neurology* 20, no. 1 (2007): 58–62.

56. R. Camicioli, D. Howieson, S. Lehman, and J. Kaye, "Talking While Walking: The Effect of a Dual Task in Aging and Alzheimer's Disease," *Neurology* 48, no. 4 (1997): 955–958.

57. F. Hamilton, L. Rochester, L. Paul, D. Rafferty, C. P. O'Leary, and J. J. Evans, "Walking and Talking: An Investigation of Cognitive–Motor Dual Tasking in Multiple Sclerosis," *Multiple Sclerosis* 15, no. 10 (2009): 1215–1227; G. Yogev, N. Giladi, C. Peretz, S. Springer, E. S. Simon, and J. M. Hausdorff, "Dual Tasking, Gait Rhythmicity, and Parkinson's Disease: Which Aspects of Gait Are Attention Demanding?" *European Journal of Neuroscience* 22, no. 5 (2005): 1248–1256.

CHAPTER 6: THE PSYCHOLOGY OF TECHNOLOGY

1. A. Toffler, *Future Shock* (New York: Random House, 1970).

2. A. Toffler, *The Third Wave* (New York: William Morrow, 1980).

3. Dr. Rosen originally discussed this concept in *Rewired: Understanding the iGeneration and the Way They Learn* (New York: Palgrave Macmillan, 2010).

4. http://www.statista.com/statistics/264810/number-of-monthly-active-facebook-users-worldwide/.

5. See, e.g., http://public.oed.com/the-oed-today/recent-updates-to-the-oed/.

6. B. Sparrow, J. Liu, and D. M. Wegner, "Google Effects on Memory: Cognitive Consequences of Having Information at Our Fingertips," *Science* 333, no. 6043 (2011): 776–778.

7. http://www.cnet.com/news/myspace-growth-continues-amid-criticism/.

8. http://mashable.com/2013/03/27/facebook-usage-survey/.

9. Pew Research Center, *U.S. Smartphone Use in 2015*, retrieved on March 2, 2016, from http://www.pewinternet.org/files/2015/03/PI_Smartphones_0401151.pdf.

10. http://www.marketingcharts.com/online/people-pick-up-their-smartphones-dozens-of-times-a-day-downtime-a-key-reason-38831/.

11. http://mashable.com/2013/07/11/smartphones-during-sex/.

12. http://www.cnn.com/2013/03/20/tech/mobile/mobile-video-bedroom/.

13. http://www.forbes.com/sites/jeffbercovici/2014/07/10/using-a-second-screen-while-watching-tv-is-now-the-norm/.

14. C. Marci, "A (Biometric) Day in the Life: Engaging across Media," paper presented at *Re:Think 2012*, New York, March 28, 2012.

15. E. Rose, "Continuous Partial Attention: Reconsidering the Role of Online Learning in the Age of Interruption," *Educational Technology Magazine: The Magazine for Managers of Change in Education* 50, no. 4 (2010): 41–46.

16. L. D. Rosen, L. M. Carrier, and N.A. Cheever, "Facebook and Texting Made Me Do It: Media-Induced Task Switching While Studying," *Computers in Human Behavior* 29, no. 3 (2013): 948–958.

17. V. M. Gonzalez and G. Mark, "Constant, Constant, Multitasking Craziness: Managing Multiple Working Spheres," in *Proceedings of the SIGCHI Conference on Human Factors in Computing Systems* (New York: ACM Press, 2004), 113–120.

18. H. A. M. Voorveld and M. van der Goot, "Age Differences in Media Multitasking: A Diary Study," *Journal of Broadcasting and Electronic Media* 57, no. 3 (2013): 392–408.

19. L. M. Carrier, N. A. Cheever, L. D. Rosen, S. Benitez, and J. Chang, "Multitasking across Generations: Multitasking Choices and Difficulty Ratings in Three Generations of Americans," *Computers in Human Behavior* 25 (2009): 483–489.

20. MarketingCharts, "College Students Own an Average of 7 Tech Devices," June 18, 2013, http://www.marketingcharts.com/wp/online/college-students-own-an-average-of-7-tech-devices-30430.

21. J. Nielsen, "F-shaped Pattern for Reading Web Content," April 17, 2006, https://www.nngroup.com/articles/f-shaped-pattern-reading-web-content/.

22. S. S. Krishnan and R. K. Sitaraman, "Video Stream Quality Impacts Viewer Behavior: Inferring Causality Using Quasi-experimental Designs," *IEEE/ACM Transactions on Networking (TON)* 21, no. 6 (2013): 2001–2014.

23. Jupiter Research, *Retail Web Site Performance*, June 2006, http://www.akamai.com/html/about/press/releases/2006/press_110606.html.

24. S. Lohr, "For Impatient Web Users, an Eye Blink Is Just Too Long to Wait," *New York Times*, February 29, 2012, http://www.nytimes.com/2012/03/01/technology/impatient-web-users-flee -slow-loading-sites.html.

25. J. Wajcman and E. Rose, "Constant Connectivity: Rethinking Interruptions at Work," *Organizational Studies* 32, no. 7 (2011): 941–961.

26. S. T. Iqbal and E. Horvitz, "Disruption and Recovery of Computing Tasks: Field Study, Analysis, and Directions," *CHI 2007: Proceedings of the SIGCHI Conference on Human Factors in Computing Systems* (New York: ACM Press, 2007), http://research.microsoft.com/EN-US/UM/PEOPLE/ horvitz/CHI_2007_Iqbal_Horvitz.pdf.

27. S. Charman-Anderson, "Breaking the Email Compulsion," *Guardian*, August 27, 2008, http:// www.guardian.co.uk/technology/2008/aug/28/email.addiction.

28. L. Marulanda-Carter and T. W. Jackson, "Effects of E-mail Addiction and Interruptions on Employees," *Journal of Systems and Information Technology* 14, no. 1 (2012): 82–94; T. Jackson, R. Dawson, and D. Wilson, "Case Study: Evaluating the Effect of Email Interruptions within the Workplace," in *Proceedings of EASE 2002: 6th International Conference on Empirical Assessment and Evaluation in Software Engineering*, https://dspace.lboro.ac.uk/dspace-jspui/bitstream/2134 /489/3/Ease%2525202002%252520Jackson.pdf.

29. M. Hair, K. V. Renaud, and J. Ramsay, "The Influence of Self-Esteem and Locus of Control on Perceived Email-related Stress," *Computers in Human Behavior* 23, no. 6 (2007): 2791–2803; K. Renaud, J. Ramsay, and M. Hair, "'You've Got E-Mail!' … Shall I Deal with It Now? Electronic Mail from the Recipient's Perspective," *International Journal of Human—Computer Interaction* 21, no. 3 (2007): 313–332.

30. MarketingCharts, "College Students Own an Average of 7 Tech Devices."

31. R. Hammer, M. Ronen, A. Sharon, T. Lankry, Y. Huberman, and V. Zamtsov, "Mobile Culture in College Lectures: Instructors' and Students' Perspectives," *Interdisciplinary Journal of E-Learning and Learning Objects* 6 (2010): 293–304; D. R. Tindell and R. W. Bohlander, "The Use and Abuse of Cell Phones and Text Messaging in the Classroom: A Survey of College Students," *College Teaching* 60, no. 1 (2012): 1–9.

32. T. Judd, "Making Sense of Multitasking: Key Behaviors," *Computers and Education* 63 (2013): 358–367.

33. C. Calderwood, P. L. Ackerman, and E. M. Conklin, "What Else Do College Students 'Do' While Studying? An Investigation of Multitasking," *Computers and Education* 75 (2014): 19–29.

34. Z. Wang and J. M. Tchernev, "The Myth of Media Multitasking: Reciprocal Dynamics of Media Multitasking, Personal Needs, and Gratifications," *Journal of Communication* 62, no. 3 (2012): 493–513.

35. Z. Wang, quoted in R. Nauert, "Multitasking Seems to Serve Emotional, not Productivity Needs," *PsychCentral*, May 1, 2012, http://psychcentral.com/news/2012/05/01/multitasking-seems -to-serve-emotional-not-productivity-needs/38057.html.

36. L. D. Rosen, L. M. Carrier, and N. A. Cheever, "Facebook and Texting Made Me Do It: Media-Induced Task-Switching While Studying," *Computers in Human Behavior* 29, no. 3 (2013): 948–958.

37. Carrier et al., "Causes, Effects, and Practicalities."

38. Carrier et al., "Multitasking across Generations."

39. Carrier et al., "Multitasking across Generations."

40. MarketingCharts, "TV Viewers and (Un)related Multi-Screening Activity: Screen Size May Count," July 29, 2013, http://www.marketingcharts.com/wp/television/tv-viewers-and-unrelated -multi-screening-activity-screen-size-may-count-35356/.

41. A. van Cauwenberge, G. Schaap, and R. van Roy, "TV No Longer Commands Our Full Attention: Effects of Second-Screen Viewing and Task Relevance on Cognitive Load and Learning from News," *Computers in Human Behavior*, 38 (2014): 100–109.

42. S. A. Basel and J. Gips, "Media Multitasking Behavior: Concurrent Television and Computer Usage," *CyberPsychology, Behavior, and Social Networking* 14, no. 9 (2011): 527–534.

43. J. Fitzgerald, "How Multi-Screen Consumers Are Changing Media Dynamics," August 28, 2012, https://www.comscore.com/Insights/Presentations-and-Whitepapers/2012/How-Multi -Screen-Consumers-Are-Changing-Media-Dynamics.

44. MarketingCharts, "4 in 5 Americans Multitask While Watching TV," March 22, 2013, http:// www.marketingcharts.com/wp/television/4-in-5-americans-multitask-while-watching-tv-28025/.

45. L. Ridley, "People Swap Devices 21 Times an Hour, Says OMD," *Brand Republic*, January 3, 2014, http://www.campaignlive.co.uk/news/1225960/.

46. H. Lindroos, "Effects of Social Presence on the Viewing Experience in a Second Screen Environment" (master's thesis, Aalto University, 2014), http://media.tkk.fi/visualmedia/publications/msc -theses/DI_H_Lindroos_2014.pdf.

47. S. Schieman and M. Young, "Who Engages in Work–Family Multitasking? A Study of Canadian and American Workers," *Social Indicators Research* 120, no. 3 (2015): 741–767, http://link .springer.com/article/10.1007/s11205-014-0609-7/fulltext.html.

48. M. Gorges, "90 Percent of Young People Wake Up with Their Smartphones," December 21, 2012, http://www.ragan.com/Main/Articles/90_percent_of_young_people_wake_up_with_their _smar_45989.aspx.

49. Bank of America, "Trends in Consumer Mobility Report" (2014), http://newsroom. bankofamerica.com/sites/bankofamerica.newshq.businesswire.com/files/press_kit/additional/ 2014_BAC_Trends_in_Consumer_Mobility.pdf.

50. MarketingCharts, "8 in 10 Smart Device Owners Use Them 'All the Time' on Vacation," April 25, 2013, http://www.marketingcharts.com/online/8-in-10-smart-device-owners-use-them-all -the-time-on-vacation-28979/.

51. Iqbal and Horvitz, "Disruption and Recovery of Computing Tasks."

CHAPTER 7: THE IMPACT OF CONSTANTLY SHIFTING OUR ATTENTION

1. L. D. Rosen, L. M. Carrier, and N. A. Cheever, "Facebook and Texting Made Me Do It: Media-Induced Task Switching While Studying," *Computers in Human Behavior* 29, no. 3 (2013): 948–958.

2. L. L. Bowman, L. E. Levine, B. M. Waite, and M. Gendron, "Can Students Really Multitask? An Experimental Study of Instant Messaging While Reading," *Computers and Education* 54 (2010): 927–931.

3. D. Prabu, J.-H. Kim, J. S. Brickman, W. Ran, and C. M. Curtis, "Mobile Phone Distraction While Studying," *New Media and Society* (2014), http://nms.sagepub.com/content/early/2014/04/22/1461444814531692.abstract.

4. For an excellent summary of the impact of a variety of devices and media uses on academic performance, see T. Judd, "Making Sense of Multitasking: The Role of Facebook," *Computers and Education* 70 (2014): 194–202.

5. G. Mark, Y. Wang, and M. Niiya, "Stress and Multitasking in Everyday College Life: An Empirical Study of Online Activity," in *CHI 2014*, 41–50; L. E. Levine, B. M. Waite, and L. L. Bowman, "Electronic Media Use, Reading, and Academic Distractibility in College Youth," *CyberPsychology and Behavior* 10, no. 4 (2007): 560–566; Bowman et al., "Can Students Really Multitask?"

6. L. M. Carrier, L. D. Rosen, N. A. Cheever, and A. F. Lim, "Causes, Effects, and Practicalities of Everyday Multitasking," *Developmental Review* 35 (2015): 64–78.

7. E. Wood, L. Zivcakova, P. Gentile, K. Archer, D. De Pasquale, and A. Nosko, "Examining the Impact of Off-Task Multi-Tasking with Technology on Real-Time Classroom Learning," *Computers and Education* 58, no. 1 (2012): 365–374.

8. J. H. Kuznekoff and S. Titsworth, "The Impact of Mobile Phone Usage on Student Learning," *Communication Education* 62, no. 3 (2013): 233–252.

9. D. E. Clayson and D. A. Haley, "An Introduction to Multitasking and Texting Prevalence and Impact on Grades and GPA in Marketing Classes," *Journal of Marketing Education* 35, no. 1 (2013): 26–40; Clayson and Haley, "An Introduction to Multitasking and Texting Prevalence"; L. Burak, "Multitasking in the University Classroom," *International Journal for the Scholarship of Teaching and Learning* 6, no. 2 (2012), http://digitalcommons.georgiasouthern.edu/ij-sotl/vol6/iss2/8; A. Lepp, J. E. Barkley, and A. C. Karpinski, "The Relationship between Cell Phone Use, Academic Performance, Anxiety, and Satisfaction with Life in College Students," *Computers in Human Behavior* 31 (2014): 343–350; R. Junco and S. R. Cotten, "No A 4 U: The Relationship between Multitasking and Academic Performance," *Computers and Education* 59, no. 2 (2012): 505–514.

10. L. D. Rosen, A. F. Lim, L. M. Carrier, and N. A. Cheever, "An Empirical Examination of the Educational Impact of Text Message-Induced Task Switching in the Classroom: Educational Implications and Strategies to Enhance Learning," *Psicologia Educativa* 17, no. 2 (2011): 163–177.

11. A. D. Froese, C. N. Carpenter, D. A. Inman, J. R. Schooley, R. B. Barnes, P. W. Brecht, and J. D. Chacon, "Effects of Classroom Cell Phone Use on Expected and Actual Learning," *College Student Journal* 46, no. 2 (2012): 323–332.

12. Lepp, Barkley, and Karpinski, "The Relationship between Cell Phone Use, Academic Performance, Anxiety, and Satisfaction."

13. M. Kalpidou, D. Costin, and J. Morris, "The Relationship between Facebook and the Well-Being of Undergraduate College Students," *Cyberpsychology, Behavior, and Social Networking* 14, no. 4 (2011): 183–189.

14. Burak, "Multitasking in the University Classroom."

15. For more information on this phenomenon, see C. N. Davidson, *Now You See It: How the Brain Science of Attention Will Transform the Way We Live, Work, and Learn* (New York: Viking Press, 2011); G. M. Slavich and P. G. Zimbardo, "Out of Mind, Out of Sight: Unexpected Scene Elements Frequently Go Unnoticed Until Primed," *Current Psychology* 32, no. 4 (2013): 310–317.

16. I. E. Hyman, S. M. Boss, B. M. Wise, K. E. McKenzie, and J. M. Caggiano, "Did You See the Unicycling Clown? Inattentional Blindness While Walking and Talking on a Cell Phone," *Applied Cognitive Psychology* 24 (2010): 597–607.

17. CBS News, "Texting While Walking, Woman Falls into Fountain," January 20, 2011, http://www.cbsnews.com/news/texting-while-walking-woman-falls-into-fountain/.

18. S. Mirsky, "Smartphone Use While Walking Is Painfully Dumb," *Scientific American* 309, no. 6 (2013), November 19, http://www.scientificamerican.com/article/smartphone-use-while-walking-is-painfully-dumb/.

19. C. H. Basch, D. Ethan, S. Rajan, and C. E. Basch, "Technology-Related Distracted Walking Behaviors in Manhattan's Most Dangerous Intersections," *Injury Prevention*, March 25, 2014, http://injuryprevention.bmj.com/content/early/2014/03/25/injuryprev-2013-041063.abstract.

20. L. L. Thompson, F. P. Rivara, R. C. Ayyagari, and B. E. Ebel, "Impact of Social and Technological Distraction on Pedestrian Crossing Behavior: An Observational Study," *Injury Prevention* 19, no. 4 (2013): 232–237.

21. N. D. Parr, C. J. Hass, and M. D. Tillman, "Cellular Phone Texting Impairs Gait in Able-Bodied Young Adults," *Journal of Applied Biomechanics* 30, no. 6 (2014): 685–688.

22. Thompson et al., "Impact of Social and Technological Distraction on Pedestrian Crossing Behavior"; Basch et al., "Technology-Related Distracted Walking Behaviors."

23. D. C. Schwebel, D. Stavrinos, K. W. Byington, T. Davis, E. E. O'Neal, and D. de Jong, "Distraction and Pedestrian Safety: How Talking on the Phone, Texting, and Listening to Music Impact Crossing the Street," *Accident Analysis and Prevention* 45 (2012): 266–271.

24. Centers for Disease Control, *Injury Prevention and Control: Motor Vehicle Safety—Distracted Driving*, http://www.cdc.gov/motorvehiclesafety/distracted_driving/.

25. National Safety Council, "National Safety Council Estimates That at Least 1.6 Million Crashes Each Year Involve Drivers Using Cell Phones and Texting" (2010), http://www.nsc.org/pages/nscestimates16millioncrashescausedbydriversusingcellphonesandtexting.aspx.

26. http://topics.nytimes.com/top/news/technology/series/driven_to_distraction/index.html.

27. D. L. Strayer, F. A. Drews, and D. J. Crouch, "A Comparison of the Cell Phone Driver and the Drunk Driver," *HFES* 48, no. 2 (2006): 381–391, http://www.psych.utah.edu/lab/appliedcognition/publications/comparison.pdf.

28. D. L. Strayer, J. M. Watson, and F. A. Drews, "Cognitive Distraction While Multitasking in the Automobile," *Psychology of Learning and Motivation-Advances in Research and Theory* 54 (2011): 29.

29. D. L. Strayer and F. A. Drews, "Cell-Phone-Induced Driver Distraction," *Current Directions in Psychological Science* 16, no. 3 (2007): 128–131.

30. J. M. Cooper, H. Ingebretsen, and D. L. Strayer, "Mental Workload of Common Voice-Based Vehicle Interactions across Six Different Vehicle Systems" (2014), https://www.aaafoundation.org/sites/default/files/Cog%20Distraction%20Phase%20IIA%20FINAL%20FTS%20FORMAT.pdf.

31. F. A. Drews, M. Pasupathi, and D. L. Strayer, "Passenger and Cell Phone Conversations in Simulated Driving," *Journal of Experimental Psychology: Applied* 14, no. 4 (2008): 392.

32. M. Madden and A. Lenhart, *Teens and Distracted Driving: Texting, Talking and Other Uses of the Cell Phone behind the Wheel* (Washington, DC: Pew Research Center's Internet & American Life Project, 2009), http://pewinternet.org/Reports/2009/Teens-and-Distracted-Driving.aspx.

33. EndDD, *End Distracted Driving Resources*, http://enddd.org/distracted-driving-resources/.

34. L. E. Levine, B. M. Waite, and L. L. Bowman, "Mobile Media Use, Multitasking, and Distractibility," *International Journal of Cyber Behavior, Psychology, and Learning* 2, no. 3 (2012): 15–29.

35. B. C. Lin, J. M. Kain, and C. Fritz, "Don't Interrupt Me! An Examination of the Relationship between Intrusions at Work and Employee Strain," *International Journal of Stress Management* 20, no. 2 (2013): 77–94.

36. V. M. Gonzalez and G. Mark, "Constant, Constant, Multitasking Craziness: Managing Multiple Working Spheres," in *Proceedings of CHI '04* (New York: ACM Press, 2004), 113–120; G. Mark, D. Gudith, and U. Klocke, "The Cost of Interrupted Work: More Speed and Stress," in *Proceedings of CHI '08* (New York: ACM Press, 2008), 107–110.

37. C. Thompson, "Meet the Life Hackers," *New York Times*, October 16, 2005, http://www.nytimes.com/2005/10/16/magazine/meet-the-life-hackers.html.

38. L. Kaufman, "Google Got It Wrong: The Open-Office Trend Is Destroying the Workplace," *Washington Post*, December 30, 2014, http://www.washingtonpost.com/posteverything/wp/2014/12/30/google-got-it-wrong-the-open-office-trend-is-destroying-the-workplace/.

39. P. K. Juneja, "Auditory Distractions in Open Office Settings: A Multi Attribute Utility Approach to Workspace Decision Making," *Dissertation Abstracts International Section A: Humanities and Social Sciences* 71, no. 11-A (2010): 3823; C. Congdon, D. Flynn, and M. Redman, "Balancing 'We' and 'Me,'" *Harvard Business Review* 92, no. 10 (2014): 50–57; J. Kim and R. de Dear, "Workspace Satisfaction: The Privacy-Communication Trade-Off in Open-Plan Offices," *Journal of Environmental Psychology* 36 (2013): 18–26.

40. A. Haapakangas, V. Hongisto, J. Hyönä, J. Kokko, and J. Keränen, "Effects of Unattended Speech on Performance and Subjective Distraction: The Role of Acoustic Design in Open-Plan Offices," *Applied Acoustics* 86 (2014): 1–16.

41. A. Seddigh, E. Berntson, C. B. Danielson, and H. Westerlund, "Concentration Requirements Modify the Effect of Office Type on Indicators of Health and Performance," *Journal of Environmental Psychology* 38 (2014): 167–174.

42. A. Shafaghat, A. Keyvanfar, H. Lamit, S. A. Mousavi, and M. Z. A. Majid, "Open Plan Office Design Features Affecting Staff's Health and Well-Being Status," *Jurnal Teknologi* 70, no. 7 (2014): 83–88.

43. J. B. Spira and J. B. Feintuch, *The Cost of Not Paying Attention: How Interruptions Impact Knowledge Worker Productivity* (September 2005), Basex, http://interruptions.net/literature/Spira-Basex05.pdf.

44. S. Turkle, *Alone Together: Why We Expect More from Technology and Less from Each Other* (New York: Basic Books, 2011).

45. A. Lenhart and M. Duggan, *Couples, the Internet, and Social Media* (Pew Research, 2014), http://www.pewinternet.org/2014/02/11/couples-the-internet-and-social-media/.

46. A. K. Przybylski and N. Weinstein, "Can You Connect with Me Now? How the Presence of Mobile Communication Technology Influences Face-to-Face Conversation Quality," *Journal of Social and Personal Relationships* 30, no. 3 (2013): 237–246.

47. S. Misra, L. Cheng, J. Genevie, and M. Yuan, "The iPhone Effect: The Quality of In-Person Social Interactions in the Presence of Mobile Devices," *Environment and Behavior* 48, no. 2 (2016): 275–298.

48. B. Thornton, A. Faires, M. Robbins, and E. Rollins, "The Mere Presence of a Cell Phone May Be Distracting: Implications for Attention and Task Performance," *Social Psychology* 45 (2014): 479–488.

49. M. Drouin, D. H. Kaiser, and D. A. Miller, "Phantom Vibrations among Undergraduates: Prevalence and Associated Psychological Characteristics," *Computers in Human Behavior* 28 (2012): 1490–1496; M. B. Rothberg, A. Arora, J. Hermann, P. St. Marie, and P. Visintainer, "Phantom Vibration Syndrome among Medical Staff: A Cross Sectional Survey," *British Medical Journal* 341, no. 12 (2010): 6914.

50. L. D. Rosen, K. Whaling, S. Rab, L. M. Carrier, and N. A. Cheever, "Is Facebook Creating 'iDisorders'? The Link between Clinical Symptoms of Psychiatric Disorders and Technology Use, Attitudes, and Anxiety," *Computers in Human Behavior* 29, no. 3 (2013): 1243–1254.

51. N. A. Cheever, L. D. Rosen, L. M. Carrier, and A. Chavez, "Out of Sight Is not Out of Mind: The Impact of Restricting Wireless Mobile Device Use on Anxiety Levels among Low, Moderate, and High Users," *Computers in Human Behavior* 37 (2014): 290–297.

52. Drouin, Kaiser, and Miller, "Phantom Vibrations among Undergraduates"; Rothberg et al., "Phantom Vibration Syndrome among Medical Staff."

53. P. A. Lewis, *The Secret World of Sleep: The Surprising Science of the Mind at Rest* (New York: Palgrave Macmillan, 2013); S. D. Sparks, "'Blue Light' May Impair Students' Sleep, Studies Say," *Education Week* 33, no. 14 (2013): 20–21.

54. S. Lemola, N. Perkinson-Gloor, S. Brand, J. F. Dewald-Kaufmann, and A. Grob, "Adolescents' Electronic Media Use at Night, Sleep Disturbance, and Depressive Symptoms in the Smartphone Age," *Journal of Youth and Adolescence* 44, no. 2 (2014): 405–418; L. Hale and S. Guan, "Screen Time and Sleep among School-Aged Children and Adolescents: A Systematic Literature Review," *Sleep Medicine Reviews* 21 (2015): 50–58.

55. S. K. Adams and T. S. Kisler, "Sleep Quality as a Mediator between Technology-Related Sleep Quality, Depression, and Anxiety," *CyberPsychology, Behavior, and Social Networking* 16, no. 1 (2013): 25–30.

56. M. Gradisar, A. R. Wolfson, A. G. Harvey, L. Hale, R. Rosenberg, and C. A. Czeisler, "The Sleep and Technology Use of Americans: Findings from the National Sleep Foundation's 2011 Sleep in America Poll," *Journal of Clinical Sleep Medicine* 9, no. 12 (2013): 1291–1299; K. A. Bartel, M. Gradisar, and P. Williamson, "Protective and Risk Factors for Adolescent Sleep: A Meta-Analytic Review," *Sleep Medicine Reviews* 21 (2015): 72–85.

57. J. Falbe, K. K. Davison, R. L. Franckle, C. Ganter, S. L. Gortmaker, L. Smith, T. Land, and E. M. Taveras, "Sleep Duration, Restfulness, and Screens in the Sleep Environment," *Pediatrics* 135, no. 2 (2015): 1–9, http://pediatrics.aappublications.org/content/early/2015/01/01/peds.2014-2306 .full.pdf.

58. A.-M. Chang, D. Aeschbach, J. F. Duffy, and C. A. Czeisler, "Evening Use of Light-Emitting eReaders Negatively Affects Sleep, Circadian Timing, and Next-Morning Alertness," *PNAS* (2014), http://www.pnas.org/content/early/2014/12/18/1418490112.full.pdf.

59. L. Rosen, L. M. Carrier, A. Miller, J. Rokkum, and Ruiz, "Sleeping with Technology: Cognitive, Affective, and Technology Usage Predictors of Sleep Problems among College Students," *Sleep Health* 2, no. 1 (2016): 49–56.

60. S. K. Adams and T. S. Kisler, "Sleep Quality as a Mediator between Technology-Related Sleep Quality, Depression, and Anxiety," *CyberPsychology, Behavior, and Social Networking* 16, no. 1 (2013): 25–30.

61. J. R. Lim, "All-Nighters Could Alter Your Memories," *Scientific American*, July 28, 2014, http://www.scientificamerican.com/article/all-nighters-could-alter-your-memories/.

62. S. J. Frenda, L. Patihis, E. F. Loftus, H. C. Lewis, and K. M. Fenn, "Sleep Deprivation and False Memories," *Psychological Science* 25, no. 9 (2014): 1674–1681.

63. A. Park, "School Should Start Later so Teens Can Sleep, Urge Doctors," *Time*, August 25, 2014, http://time.com/3162265/school-should-start-later-so-teens-can-sleep-urge-doctors; Adolescent Sleep Working Group, Committee on Adolescence, and Council of School Health, "School Start Times for Adolescents," *Pediatrics* 134, no. 3 (2014): 642–649.

64. Mayo Clinic, "Are Smartphones Disrupting Your Sleep? Mayo Clinic Examines the Question," June 3, 2013, http://newsnetwork.mayoclinic.org/discussion/are-smartphones-disrupting-your-sleep-mayo-clinic-study-examines-the-question/?mc_id=youtube.

65. K. Lanaj, R. E. Johnson, and C. M. Barnes, "Beginning the Workday yet Already Depleted? Consequences of Late-Night Smartphone Use and Sleep," *Organizational Behavior and Human Decision Processes* 124, no. 1 (2014): 11–23.

66. K. Custers and J. Van den Bulck, "Television Viewing, Internet Use, and Self-Reported Bedtime and Rise Time in Adults: Implications for Sleep Hygiene Recommendations from an Exploratory Cross-Sectional Study," *Behavioral Sleep Medicine* 10, no. 2 (2012): 96–105.

CHAPTER 8: THE IMPACT OF TECHNOLOGY ON DIVERSE POPULATIONS

1. L. D. Rosen, *iDisorder: Understanding Our Obsession with Technology and Overcoming Its Hold on Us* (New York: Palgrave-Macmillan, 2012); L. D. Rosen and J. Lara-Ruiz, "Similarities and Differences in Workplace, Personal, and Technology-Related Values, Beliefs, and Attitudes across Five Generations of Americans," in *The Handbook of Psychology, Technology, and Society*, ed. L. D. Rosen, N. A. Cheever, and L. M. Carrier (Hoboken, NJ: Wiley-Blackwell, 2015).

2. L. M. Carrier, N. A. Cheever, L. D. Rosen, S. Benitez, and J. Chang, "Multitasking across Generations: Multitasking Choices and Difficulty Ratings in Three Generations of Americans," *Computers in Human Behavior* 25, no. 2 (2009): 483–489.

3. M. Prensky, "Digital Natives, Digital Immigrants," *On the Horizon* 9 (2001): 1–6.

4. K. L. Mills, F. Lalonde, L. S. Clasen, J. N. Giedd, and S. J. Blakemore, "Developmental Changes in the Structure of the Social Brain in Late Childhood and Adolescence," *Social Cognitive and Affective Neuroscience*, 9, no. 1 (2014): 123–131.

5. J. Mishra, J. A. Anguera, D. A. Ziegler, and A. Gazzaley, "A Cognitive Framework for Understanding and Improving Interference Resolution in the Brain," *Progress in Brain Research* 207 (2013): 351–377.

6. D. A. Christakis, F. J. Zimmerman, D. L. DiGiuseppe, and C. A. McCarty, "Early Television Exposure and Subsequent Attentional Problems in Children," *Pediatrics* 113, no. 4 (2004): 708–713; American Academy of Pediatrics, *Media and Children* (n.d.), http://www.aap.org/en-us/advocacy-and-policy/aap-health-initiatives/Pages/Media-and-Children.aspx.

7. eMarketer, "Most US Children Use the Internet at Least Daily," April 28, 2014, http://www.emarketer.com/Article/Most-US-Children-Use-Internet-Least-Daily/1010789.

8. A. L. Gutnick, M. Robb, L. Takeuchi, and J. Kotler, *Always Connected: The New Digital Media Habits of Young Children* (New York: The Joan Ganz Cooney Center at Sesame Workshop, 2010).

9. C. Wallis, *The Impacts of Media Multitasking on Children's Learning and Development: Report from a Research Seminar* (The Joan Ganz Cooney Center and Stanford University, 2010); V. Rideout, U. G. Foehr, and D. Roberts, *Generation M2: Media in the Lives of 8- to 18-Year-Olds* (Menlo Park, CA: Henry J. Kaiser Family Foundation, 2010).

10. Commonsense Media, "Entertainment Media Diets of Children and Adolescents May Impact Learning," November 1, 2012, https://www.commonsensemedia.org/about-us/news/press-releases/entertainment-media-diets-of-children-and-adolescents-may-impact; K. Purcell, L. Rainie, A. Heaps, J. Buchanan, L. Friedrich, A. Jacklin, C. Chen, and K. Zickuhr, *How Teens Do Research in the Digital World*, November 1, 2012, http://www.pewinternet.org/2012/11/01/how-teens-do-research-in-the-digital-world/.

11. Gutnick et al., *Always Connected*.

12. M. Honan, "Are Touchscreens Melting Your Kid's Brain?" *Wired*, April 15, 2014, http://www.wired.com/2014/04/children-and-touch-screens/.

13. L. D. Rosen, A. F. Lim, J. Felt, L. M. Carrier, N. A. Cheever, J. M. Lara-Ruiz, J. S. Mendoza, and J. Rokkum, "Media and Technology Use Predicts Ill-Being among Children, Preteens, and Teenagers Independent of the Negative Health Impacts of Exercise and Eating Habits," *Computers in Human Behavior* 35 (2014): 364–375.

14. A. S. Page, A. R. Cooper, P. Griew, and R. Jago, "Children's Screen Viewing Is Related to Psychological Difficulties Irrespective of Physical Activity," *Pediatrics* 126, no. 5 (2010): e1011–e1017; A. Parkes, H. Sweeting, D. Wight, and M. Henderson, "Do Television and Electronic Games Predict Children's Psychosocial Adjustment? Longitudinal Research Using the UK Millennium Cohort Study," *Archives of Disease in Childhood* 98, no. 5 (2013): 341–348; T. Hinkley, V. Verbestel, W. Ahrens, L. Lissner, D. Molnár, L. A. Moreno, I. Pigeot, et al., "Early Childhood Electronic Media Use as a Predictor of Poorer Well-Being: A Prospective Cohort Study," *JAMA Pediatrics* 168, no. 5 (2014): 485–492; R. Pea, C. Nass, L. Meheula, M. Rance, A. Kumar, H. Bamford, M. Nass, et al., "Media Use, Face-to-Face Communication, Media Multitasking, and Social Well-Being among 8- to 12-Year-Old Girls," *Developmental Psychology* 48, no. 2 (2012): 327; D. A. Gentile, E. L. Swing, C. G. Lim, and A. Khoo, "Video Game Playing, Attention Problems, and Impulsiveness: Evidence of Bidirectional Causality," *Psychology of Popular Media Culture* 1, no. 1 (2012): 62; K. Subrahmanyam, R. E. Kraut, P. M. Greenfield, and E. F. Gross,

"The Impact of Home Computer Use on Children's Activities and Development," *Children and Computer Technology* 10, no. 2 (2000): 123–144; L. S. Pagani, C. Fitzpatrick, T. A. Barnett, and E. Dubow, "Prospective Associations between Early Childhood Television Exposure and Academic, Psychosocial, and Physical Well-Being by Middle Childhood," *Archives of Pediatrics and Adolescent Medicine* 164, no. 5 (2010): 425–431.

15. National Sleep Foundation, *2014 Sleep in America Poll: Sleep in the Modern Family, Summary of Findings*, http://sleepfoundation.org/sites/default/files/2014-NSF-Sleep-in-America-poll-summary -of-findings---FINAL-Updated-3-26-14-.pdf (last accessed January 28, 2016); Harvard Medical School Family Health Guide, *Repaying Your Sleep Debt*, http://www.health.harvard.edu/fhg/ updates/Repaying-your-sleep-debt.shtml (last accessed January 82, 2016).

16. N. Cain and M. Gradisar, "Electronic Media Use and Sleep in School-Aged Children and Adolescents: A Review," *Sleep Medicine* 11, no. 8 (2010): 735–742.

17. A. A. K. Morsy and N. S. Shalaby, "The Use of Technology by University Adolescent Students and Its Relation to Attention, Sleep, and Academic Achievement," *Journal of American Science* 8, no. 1 (2012): 264–270; L. S. Foley, R. Maddison, Y. Jiang, S. Marsh, T. Olds, and K. Ridley, "Presleep Activities and Time of Sleep Onset in Children," *Pediatrics* 131, no. 2 (2013): 276–282; E. J. Paavonen, M. Pennonen, M. Roine, S. Valkonen, and A. R. Lahikainen, "TV Exposure Associated with Sleep Disturbances in 5- to 6-Year-Old Children," *Journal of Sleep Research* 15, no. 2 (2006): 154–161; T. Nuutinen, C. Ray, and E. Roos, "Do Computer Use, TV Viewing, and the Presence of the Media in the Bedroom Predict School-Aged Children's Sleep Habits in a Longitudinal Study," *BMC Public Health* 13, no. 1 (2013): 684.

18. S. J. Blakemore and T. W. Robbins, "Decision-Making in the Adolescent Brain," *Nature Neuroscience* 15, no. 9 (2012): 1184–1191.

19. J. K. Mullen, "The Impact of Computer Use on Employee Performance in High-Trust Professions: Re-examining Selection Criteria in the Internet Age," *Journal of Applied Social Psychology* 41, no. 8 (2011): 2009–2043.

20. Rosen, *iDisorder*.

21. U. G. Foehr, *Media Multitasking among American Youth: Prevalence, Predictors, and Pairings: Report* (Menlo Park, CA: Kaiser Family Foundation, 2006), http://www.kff.org/entmedia/upload/ 7592.pdf.

22. Rosen and Lara-Ruiz, "Similarities and Differences in Workplace, Personal, and Technology-Related Values."

23. Nielsen.com, "New Mobile Obsession U.S. Teens Triple Data Usage," December 15, 2011, http://www.nielsen.com/us/en/insights/news/2011/new-mobile-obsession-u-s-teens-triple-data -usage.html.

24. A. Smith, "Older Adults and Technology Use," April 3, 2014, http://www.pewinternet.org/ 2014/04/03/older-adults-and-technology-use/.

25. H. A. M. Voorveld and M. van der Goot, "Age Differences in Media Multitasking: A Diary Study," *Journal of Broadcasting and Electronic Media* 57, no. 3 (2013): 392–408.

26. L. D. Rosen and J. Lara-Ruiz, "Similarities and Differences in Workplace, Personal and Technology-Related Values, Beliefs, and Attitudes across Five Generations of Americans," in *The Handbook of Psychology, Technology, and Society*, ed. L. D. Rosen, N. A. Cheever, and L. M. Carrier (Hoboken, NJ: Wiley-Blackwell, 2015), 20–55.

27. J. T. E. Richardson and A. Jelfs, "Access and Attitudes to Digital Technologies Across the Lifespan," in *The Handbook of Psychology, Technology, and Society*, 89–104; K. Magsamen-Conrad, J. Dowd, M. Abuljadail, S. Alsulaiman, and A. Shareefi, "Life-Span Differences in the Uses and Gratifications of Tablets: Implications for Older Adults," *Computers in Human Behavior* 52 (2015): 96–106.

28. A. M. Kueider, J. M. Parisi, A. L. Gross, and G. W. Rebok, "Computerized Cognitive Training with Older Adults: A Systematic Review," *PLoS ONE* 7, no. 7 (2012): e40588.

29. S. Reaves, B. Graham, J. Grahn, P. Rabinnifard and A. Duarte, "Turn Off the Music! Music Impairs Visual Associative Memory Performance in Older Adults," *Gerontologist* (2015), https://gerontologist.oxfordjournals.org/content/early/2015/01/28/geront.gnu113.full.

30. N. Takeuchi, T. Mori, Y. Suzukamo, N. Tanaka, and S. I. Izumi, "Parallel Processing of Cognitive and Physical Demands in Left and Right Prefrontal Cortices during Smartphone Use While Walking," *BMC Neuroscience* 17, no. 1 (2016): 1.

31. L. L. Thompson, F. P. Rivara, R. C. Ayyagari, and B. E. Ebel, "Impact of Social and Technological Distraction on Pedestrian Crossing Behaviour: An Observational Study," *Injury Prevention* 19, no. 4 (2013): 232–237.

32. E. K. Vernon, G. M. Babulal, G. Head, D. Carr, N. Ghoshal, P. P. Barco, J. C. Morris, and C. M. Roe, "Adults Aged 65 and Older Use Potentially Distracting Electronic Devices While Driving," *Journal of American Geriatrics Society* 63, no. 6 (2015): 1251–1254.

33. World Health Organization, "Mobile Phone Use: A Growing Problem of Driver Distraction" (Geneva: WHO, 2011), retrieved on February 25, 2016 from http://apps.who.int/iris/bitstream/10665/44494/1/9789241500890_eng.pdf.

34. S. M. Ravizza and R. E. Salo, "Task Switching in Psychiatric Disorders," in *Task Switching*, ed. J. Grange and G. Houghton (Oxford: Oxford University Press, 2014), 300–331.

35. D. Getahun, S. J. Jacobsen, M. J. Fassett, W. Chen, K. Demissie, and G. G. Rhoads, "Recent Trends in Childhood Attention-Deficit/Hyperactivity Disorder," *JAMA Pediatrics* 167, no. 3 (2013): 282–288.

36. S. W. Nikkelen, P. M. Valkenburg, M. Huizinga, and G. J. Bushman, "Media Use and ADHD-Related Behaviors in Children and Adolescents: A Meta-Analysis," *Developmental Psychology* 50, no. 9 (2014): 2228–2241.

37. Ravizza and Salo, "Task Switching in Psychiatric Disorders."

38. J. B. Ewen, J. S. Moher, B. M. Lakshmanan, M. Ryan, P. Xavier, N. E. Crone, M. B. Denckla, et al., "Multiple Task Interference Is Greater in Children with ADHD," *Developmental Neuropsychology* 37, no. 2 (2012): 119–133.

39. Dana Foundation, *ADHD, Multi-Tasking, and Reading*, May 7, 2012, http://danablog.org/2012/05/07/adhd-reading-multitasking/.

40. S. Siklos and K. A. Kerns, "Assessing Multitasking in Children with ADHD Using a Modified Six Elements Test," *Archives of Clinical Neuropsychology* 19, no. 3 (2004): 347–361. For an excellent summary of how the ADHD brain processes information differently than others, see J. Mishra, J. A. Anguera, D. A. Ziegler, and A. Gazzaley, "A Cognitive Framework for Understanding and Improving Interference Resolution in the Brain," *Progress in Brain Research* 207 (2013) 351–377.

41. M. Narad, A. A. Garner, A. A. Brassell, D. Saxby, T. N. Antonini, K. M. O'Brien, L. Tamm, G. Matthews, and J. N. Epstein, "Impact of Distraction on the Driving Performance of Adolescents With and Without Attention-Deficit/Hyperactivity Disorder," *JAMA Pediatrics*, 167, no. 10 (2013): 933–938.

42. G. S. O'Keeffe and K. Clarke-Pearson, "The Impact of Social Media on Children, Adolescents, and Families," *Pediatrics* 127, no. 4 (2011): 800–804.

43. L. D. Rosen, K. Whaling, S. Rab, L. M. Carrier, and N. A. Cheever, "Is Facebook Creating 'iDisorders'? The Link between Clinical Symptoms of Psychiatric Disorders and Technology Use, Attitudes, and Anxiety," *Computers in Human Behavior*, 29, no. 3 (2013): 1243–1254.

44. E. B. Thorsteinsson and L. Davey, "Adolescents' Compulsive Internet Use and Depression: A Longitudinal Study," *Open Journal of Depression* 3 (2014): 13.

45. L. L. Lou, Z. Yan, A. Nickerson, and R. McMorris, "An Examination of the Reciprocal Relationship of Loneliness and Facebook Use among First-Year College Students," *Journal of Educational Computing Research* 46, no. 1 (2012): 105–117; K. B. Wright, J. Rosenberg, N. Egbert, N. A. Ploeger, D. R. Bernard, and S. King, "Communication Competence, Social Support, and Depression among College Students: A Model of Facebook and Face-to-Face Support Network Influence," *Journal of Health Communication*, 18, no. 1 (2013): 41–57.

46. F. Große Deters and M. R. Mehl, "Does Posting Facebook Status Updates Increase or Decrease Loneliness? An Online Social Networking Experiment," *Social Psychological and Personality Science* 4, no. 5 (2013): 579–586.

47. S. R. Tortolero, M. F. Peskin, E. F. Baumler, P. M. Cuccaro, M. N. Elliott, S. L. Davies, T. H. Lewis, et al., "Daily Violent Video Game Playing and Depression in Preadolescent Youth," *Cyberpsychology, Behavior, and Social Networking* 17, no. 9 (2014): 609–615; I. Pantic, A. Damjanovic, J. Todorovic, D. Topalovic, D. Bojovic-Jovic, S. Ristic, and S. Pantic, "Association between Online Social Networking and Depression in High School Students: Behavioral Physiology Viewpoint," *Psychiatria Danubina* 24, no. 1 (2012): 90–93; S. Thomée, A. Härenstam, and M. Hagberg, "Computer Use and Stress, Sleep Disturbances, and Symptoms of Depression among Young Adults—A Prospective Cohort Study," *BMC Psychiatry*, 12, no. 1 (2012): 176; M. A. Moreno, L. A. Jelenchick, K. G. Egan, E. Cox, H. Young, K. E. Gannon, and T. Becker, "Feeling Bad on Facebook: Depression Disclosures by College Students on a Social Networking Site," *Depression and Anxiety* 28, no. 6 (2011): 447–455; S. Thomée, M. Eklöf, E. Gustafsson, R. Nilsson, and M. Hagberg, "Prevalence of Perceived Stress, Symptoms of Depression, and Sleep Disturbances in Relation to Information and Communication Technology (ICT) Use among Young Adults—An Explorative Prospective Study," *Computers in Human Behavior* 23, no. 3 (2007): 1300–1321; Y. Al-Saggaf and S. Nielsen, "Self-Disclosure on Facebook among Female Users and Its Relationship to Feelings of Loneliness," *Computers in Human Behavior* 36 (2014): 460–468; C. M. Morrison and H. Gore, "The Relationship between Excessive Internet Use and Depression: A Questionnaire-Based Study of 1,319 Young People and Adults," *Psychopathology* 43, no. 2 (2010): 121–126; D. A. Christakis, M. M. Moreno, L. Jelenchick, M. Myaing, and C. Zhou, "Problematic Internet Usage in US College Students: A Pilot Study," *BMC Medicine* 9, no. 1 (2011): 77.

48. H. Song, A. Zmyslinski-Seelig, J. Kim, A. Drent, A. Victor, K. Omori, and M. Allen, "Does Facebook Make You Lonely? A Meta Analysis," *Computers in Human Behavior* 36 (2014): 446–452.

49. Ravizza and Salo, "Task Switching in Psychiatric Disorders."

50. R. Kotikalapudi, S. Chellappan, F. Montgomery, D. Wunsch, and K. Lutzen, "Associating Depressive Symptoms in College Students with Internet Usage Using Real Internet Data," *IEEE Technology and Society Magazine* 31, no. 4 (2012): 73–80.

51. A. D. Kramer, J. E. Guillory, and J. T. Hancock, "Experimental Evidence of Massive-Scale Emotional Contagion through Social Networks," *Proceedings of the National Academy of Sciences* 111, no. 24 (2014): 8788–8790.

52. M. W. Becker, R. Alzahabi, and C. J. Hopwood, "Media Multitasking Is Associated with Symptoms of Depression and Social Anxiety," *Cyberpsychology, Behavior, and Social Networking* 16, no. 2 (2013): 132–135.

53. Rosen et al., "Is Facebook Creating 'iDisorders'?"

54. G. T. Waldhauser, M. Johansson, M. Bäckström, and A. Mecklinger, "Trait Anxiety, Working Memory Capacity, and the Effectiveness of Memory Suppression," *Scandinavian Journal of Psychology* 52, no. 1 (2011): 21–27; L. Visu-Petra, M. Miclea, and G. Visu-Petra, "Individual Differences in Anxiety and Executive Functioning: A Multidimensional View," *International Journal of Psychology* 48, no. 4 (2011): 649–659.

55. J. M. Twenge and W. K. Campbell, *The Narcissism Epidemic: Living in the Age of Entitlement* (New York: Free Press, 2009).

56. Rosen, et al., "Is Facebook Creating 'iDisorders'?"

57. For more research and information on the impact of technology on people suffering from narcissistic personality disorder, see Rosen, *iDisorder*.

58. Ravizza and Salo, "Task Switching in Psychiatric Disorders."

59. G. Rajendran, A. S. Law, R. H. Logie, M. Van Der Meulen, D. Fraser, and M. Corley, "Investigating Multitasking in High-Functioning Adolescents with Autism Spectrum Disorders Using the Virtual Errands Task," *Journal of Autism and Developmental Disorders* 41, no. 11 (2011): 1445–1454.

60. M. L. Gonzalez-Gadea, S. Baez, T. Torralva, F. X. Castellanos, A. Rattazzi, V. Bein, K. Rogg, F. Manes, and A. Ibanez, "Cognitive Variability in Adults with ADHD and AS: Disentangling the Roles of Executive Functions and Social Cognition," *Research in Developmental Disabilities* 34, no. 2 (2013): 817–830.

61. M. O. Mazurek and C. Wenstrup, "Television, Video Game, and Social Media Use among Children with ASD and Typically Developing Siblings," *Journal of Autism and Developmental Disorders* 43, no. 6 (2013): 1258–1271.

CHAPTER 9: WHY DO WE INTERRUPT OURSELVES?

1. H. Simon, quoted in S. Anderson, "In Defense of Distraction: Twitter, Adderall, Lifehacking, Mindful Jogging, Power Browsing, Obama's BlackBerry, and the Benefits of Overstimulation," *New York Magazine*, May 17, 2009, http://nymag.com/news/features/56793/.

2. Anderson, "In Defense of Distraction."

3. P. Pirolli and S. Card, "Information Foraging," *Psychological Review* 106, no. 4 (1999): 643–675.

4. L. Yeykelis, J. J. Cummings, and B. Reeves, "Multitasking on a Single Device: Arousal and the Frequency, Anticipation, and Prediction of Switching between Media Content on a Computer," *Journal of Communication* 64 (2014): 167–192.

5. J. D. Eastwood, A. Frischen, M. J. Fenske, and D. Smilek, "The Unengaged Mind Defining Boredom in Terms of Attention," *Perspectives on Psychological Science* 7, no. 5 (2012): 482–495.

6. K. Weir, "Never a Dull Moment: Things Get Interesting When Psychologists Take a Closer Look at Boredom," *Monitor on Psychology* 44, no. 7 (2013): 52.

7. W. Miklaus and S. Vodanovich, "The Essence of Boredom," *Psychological Record* 43 (1993): 3–12.

8. J. M. Barbalet, "Boredom and Social Meaning," *British Journal of Sociology* 50, no. 4 (1999): 631–646.

9. E. Fromm, *The Anatomy of Human Destructiveness* (New York: Holt McDougal, 1973).

10. "Tech-or-Treat: Consumers Are Sweet on Mobile Apps," *Nielsen.com*, October 20, 2014, http://www.nielsen.com/us/en/insights/news/2014/tech-or-treat-consumers-are-sweet-on-mobile-apps.html.

11. "Downtime? Half of UK Smartphone Owners Prefer to Check Their Devices," *MarketingCharts.com*, October 18, 2013, http://www.marketingcharts.com/online/downtime-half-of-uk-smartphone-owners-prefer-to-check-their-devices-37529/.

12. "Boredom Said to Spur Video Sharing among Smartphone Owners," *MarketingCharts.com*, August 1, 2013, http://www.marketingcharts.com/online/boredom-said-to-spur-video-sharing-among-smartphone-owners-35474/.

13. L. D. Rosen, L. M. Carrier, and N. A. Cheever, "Facebook and Texting Made Me Do It: Media-Induced Task-Switching While Studying," *Computers in Human Behavior* 29, no. 3 (2013): 948–958.

14. A. Lepp, J. Li, J. E. Barkley, and S. Salehi-Esfahani, "Exploring the Relationships between College Students' Cell Phone Use, Personality, and Leisure," *Computers in Human Behavior* 43 (2015): 210–219; A. Lepp and J. E. Barkley, "Cell Phone Use as Leisure: Activities, Motivations, and Affective Experiences," in *Book of Abstracts for the Leisure Research Symposium of the National Recreation and Parks Association's Annual Congress* (Ashburn, VA: National Recreation and Park Association, 2014), 154–156, http://www.academyofleisuresciences.com/sites/default/files/2014%20LRS%20Book%20of%20Abstracts.pdf.

15. J. A. Danckert and A. A. A. Allman, "Time Flies When You're Having Fun: Temporal Estimation and the Experience of Boredom," *Brain and Cognition* 59, no. 3 (2005): 236–245.

16. Weir, "Never a Dull Moment."

17. Eastwood et al., "The Unengaged Mind."

18. D. Gross, "Have Smartphones Killed Boredom (and Is That Good)?" *CNN.com*, September 26, 2012, http://www.cnn.com/2012/09/25/tech/mobile/oms-smartphones-boredom/.

19. National Institute of Mental Health, "Any Anxiety Disorder Among Adults," http://www.nimh.nih.gov/health/statistics/prevalence/any-anxiety-disorder-among-adults.shtml; A. V. Horwitz and J. C. Wakefield, *All We Have to Fear: Psychiatry's Transformation of Natural Anxieties into Mental Disorders* (Oxford: Oxford University Press, 2012).

20. L. D. Rosen, K. Whaling, S. Rab, L. M. Carrier, and N. A. Cheever, "Is Facebook Creating 'iDisorders'? The Link between Clinical Symptoms of Psychiatric Disorders and Technology Use, Attitudes, and Anxiety," *Computers in Human Behavior* 29 (2013): 1243–1254.

21. C. Taylor, "For Millennials, Social Media Is not All Fun and Games," *Gigaom.com*, April 29, 2011, http://gigaom.com/2011/04/29/millennial-mtv-study/; A. K. Przybylski, K. Murayama, C. R. DeHaan, and V. Gladwell, "Motivational, Emotional, and Behavioral Correlates of Fear of Missing Out," *Computers in Human Behavior* 29, no. 4 (2013): 1841–1848.

22. Przybylski et al., "Motivational, Emotional, and Behavioral Correlates."

23. J. Loechner, "Fear of Missing Out Drives Social Media Use," *Mediapost.com*, August 7, 2012, http://www.mediapost.com/publications/article/180230/fear-of-missing-out-drives-social-media -use.html.

24. Jeff Tingsley, quoted in N. Smith, "Social Media 'Addiction' Is Marketer's Best Friend," *Business News Daily*, August 1, 2012, http://www.businessnewsdaily.com/2933-consumers-addicted-social -media-fear-missing-out.html.

25. Rosen et al., "Is Facebook Creating 'iDisorders'?"

26. "Nomophobia, the Fear of Not Having a Mobile Phone, Hits Record Numbers," *Australian*, June 2, 2013, http://www.news.com.au/technology/nomophobia-the-fear-of-not-having-a-mobile- phone-hits-record-numbers/story-e6frfro0-1226655033189.

27. S. A. Kelly, "Are You Afraid of Mobile Phone Separation?" *Mashable.com*, July 13, 2012, http:// mashable.com/2012/07/13/nomophobia-infographic/.

28. S. A. Kelly, "Afraid of Losing Your Phone? You May Have Nomophobia Like Half the Popula- tion," *Mashable.com*, February 21, 2012, http://mashable.com/2012/02/21/nomophobia/.

29. N. A. Cheever, L. D. Rosen, L. M. Carrier, and A. Chavez, "Out of Sight Is Not Out of Mind: The Impact of Restricting Wireless Mobile Device Use on Anxiety among Low, Moderate, and High Users," *Computers in Human Behavior* 37 (2014): 290–297.

30. R. B. Clayton, G. Leshner, and A. Almond, "The Extended iSelf: The Impact of iPhone Separa- tion on Cognition, Emotion, and Physiology," *Journal of Computer-Mediated Communication* 20, no. 2 (2015): 119–135.

31. A. Lepp, J. E. Barkley, and A. C. Karpinski, "The Relationship between Cell Phone Use, Aca- demic Performance, Anxiety, and Satisfaction with Life in College Students," *Computers in Human Behavior* 31 (2014): 343–350; S. Thomée, M. Eklöf, E. Gustafsson, R. Nilsson, and M. Hagberg, "Prevalence of Perceived Stress, Symptoms of Depression, and Sleep Disturbances in Relation to Information and Communication Technology (ICT) Use among Young Adults—An Explorative Prospective Study," *Computers in Human Behavior*, 23, no. 3 (2007): 1300–1321.

32. J. Bennett, "Bubbles Carry a Lot of Weight: Texting Anxiety Caused by Little Bubbles," *New York Times*, August 29, 2014, http://www.nytimes.com/2014/08/31/fashion/texting-anxiety -caused-by-little-bubbles.html.

33. C. Doctorow, "Writing in the Age of Distraction," *Locus Magazine*, January 2009, http://www .locusmag.com/Features/2009/01/cory-doctorow-writing-in-age-of.html.

34. D. Sanbonmatsu, D. Strayer, and N. Medeiros-Ward, "Who Multi-Tasks and Why? Multi-Task- ing Ability, Perceived Multi-Tasking Ability, Impulsivity, and Sensation Seeking," *PLoS ONE* 8, no. 1 (2013): e54402.

35. E. Ophir, C. Nass, and A. D. Wagner, "Cognitive Control in Media Multitaskers," *Proceedings of the National Academy of Sciences of the United States of America* 106 (2009): 15583–15587.

36. J. R. Finley, A. S. Benjamin, and J. S. McCarley, "Metacognition of Multitasking: How Well Do We Predict the Costs of Divided Attention?" *Journal of Experimental Psychology: Applied* (February 3, 2014), advance online publication, http://dx.doi.org/10.1037/xap0000010.

37. W. J. Horrey and M. F. Lesch, "Driver-Initiated Distractions: Examining Strategic Adaptation for In-Vehicle Task Initiation," *Accident Analysis and Prevention* 41 (2009): 115–122.

38. For an interesting take on the impact of text messaging while driving, see M. Richtel, *A Deadly Wandering: A Tale of Tragedy and Redemption in the Age of Attention* (New York: HarperCollins, 2014).

39. S. T. Iqbal and E. Horvitz, "Disruption and Recovery of Computing Tasks: Field Study, Analysis, and Directions," in *Proceedings of CHI 2007* (New York: ACM Press, 2007), 677–686.

40. Finley, Benjamin, and McCarley, "Metacognition of Multitasking."

CHAPTER 10: BOOSTING CONTROL

1. W. James, *The Principles of Psychology* (New York: Dover, 1890).

2. Y. Stern, "Cognitive Reserve," *Neuropsychologia* 47, no. 10 (2009): 2015–2028.

3. Y. Stern, "Cognitive Reserve and Alzheimer Disease," *Alzheimer Disease and Associated Disorders* 20, no. 2 (2006): 112–117.

4. J. Dewey, *Experience and Education* (New York: Kappa Delta Pi, 1938), 1–5.

5. On the connection between cognitive control and academic performance, see, e.g., J. A. Welsh, R. L. Nix, C. Blair, K. L. Bierman, and K. E. Nelson, "The Development of Cognitive Skills and Gains in Academic School Readiness for Children from Low-Income Families," *Journal of Educational Psychology* 102, no. 1 (2010), 43–53; C. Blair and R. P. Razza, "Relating Effortful Control, Executive Function, and False Belief Understanding to Emerging Math and Literacy Ability in Kindergarten," *Child Development* 78, no. 2 (2007): 647–663; L. Visu-Petra, L. Cheie, O. Benga, and M. Miclea, "Cognitive Control Goes to School: The Impact of Executive Functions on Academic Performance," *Procedia-Social and Behavioral Sciences* 11 (2011): 240–244.

6. D. P. Baker, D. Salinas, and P. J. Eslinger, "An Envisioned Bridge: Schooling as a Neurocognitive Developmental Institution," *Developmental Cognitive Neuroscience* 2 (2012): S6–S17.

7. C. Blair, D. Gamson, S. Thorne, and D. Baker, "Rising Mean IQ: Cognitive Demand of Mathematics Education for Young Children, Population Exposure to Formal Schooling, and the Neurobiology of the Prefrontal Cortex," *Intelligence* 33, no. 1 (2005): 93–106.

8. E. Peters, D. P. Baker, N. F. Dieckmann, J. Leon, and J. Collins, "Explaining the Effect of Education on Health: A Field Study in Ghana," *Psychological Science* 21, no. 10 (2010): 1369–1376; D. P. Baker, D. Salinas, and P. J. Eslinger, "An Envisioned Bridge: Schooling as a Neurocognitive Developmental Institution," *Developmental Cognitive Neuroscience* 2 (2012): S6–S17.

9. A. Diamond, W. S. Barnett, J. Thomas, and S. Munro, "Preschool Program Improves Cognitive Control," *Science* 318, no. 5855 (2007): 1387–1388.

10. A. R. Luria, *The Higher Cortical Functions in Man* (New York: Basic Books, 1966); L. S. Vygotsky, *Mind in Society: The Development of Higher Psychological Processes* (Cambridge, MA: Harvard University Press, 1978).

11. http://www.ncbi.nlm.nih.gov/pmc/articles/PMC2174918/.

12. J. Gu, C. Strauss, R. Bond, and K. Cavanagh, "How Do Mindfulness-Based Cognitive Therapy and Mindfulness-Based Stress Reduction Improve Mental Health and Wellbeing? A Systematic Review and Meta-Analysis of Meditation Studies," *Clinical Psychology Review* 37 (2015): 1–12.

13. A. Chiesa, R. Calati, and A. Serretti, "Does Mindfulness Training Improve Cognitive Abilities? A Systematic Review of Neuropsychological Findings," *Clinical Psychology Review* 31, no. 3 (2011): 449–464; H. A. Slagter, R. J. Davidson, and A. Lutz, "Mental Training as a Tool in the Neuroscientific Study of Brain and Cognitive Plasticity," *Frontiers in Human Neuroscience* 5 (2011): 17.

14. A. P. Jha, J. Krompinger, and M. J. Baime, "Mindfulness Training Modifies Subsystems of Attention," *Cognitive, Affective, and Behavioral Neuroscience* 7, no. 2 (2007): 109–119.

15. K. A. MacLean, E. Ferrer, S. R. Aichele, D. A. Bridwell, A. P. Zanesco, T. L. Jacobs, et al., "Intensive Meditation Training Improves Perceptual Discrimination and Sustained Attention," *Psychological science* 21, no. 6 (2010): 829–839. A. Lutz, H. A. Slagter, N. B. Rawlings, A. D. Francis, L. L. Greischar, and R. J. Davidson, "Mental Training Enhances Attentional Stability: Neural and Behavioral Evidence," *Journal of Neuroscience* 29, no. 42 (2009): 13418–13427; H. A. Slagter, A. Lutz, L. L. Greischar, S. Nieuwenhuis, and R. J. Davidson, "Theta Phase Synchrony and Conscious Target Perception: Impact of Intensive Mental Training," *Journal of Cognitive Neuroscience*, 21, no. 8 (2009): 1536–1549; H. A. Slagter, A. Lutz, L. L. Greischar, A. D. Francis, S. Nieuwenhuis, J. M. Davis, and R. J. Davidson, "Mental Training Affects Distribution of Limited Brain Resources," *PLoS Biology* 5, no. 6 (2007): e138; M. D. Mrazek, M. S. Franklin, D. T. Phillips, B. Baird, and J. W. Schooler, "Mindfulness Training Improves Working Memory Capacity and GRE Performance While Reducing Mind Wandering," *Psychological Science* 24, no. 5 (2013): 776–781; A. P. Jha, E. A. Stanley, A. Kiyonaga, L. Wong, and L. Gelfand, "Examining the Protective Effects of Mindfulness Training on Working Memory Capacity and Affective Experience," *Emotion* 10, no. 1 (2010): 54–64.

16. M. D. Mrazek, M. S Franklin, D. T. Phillips, B. Baird, and J. W. Schooler, "Mindfulness Training Improves Working Memory Capacity and GRE Performance While Reducing Mind Wandering," *Psychological Science* 24, no. 5 (2013): 776–781.

17. A. Fernandez, "The Business and Ethics of the Brain Fitness Boom," *Generations* 35, no. 2 (2011): 63–69.

18. A. Lampit, H. Hallock, and M. Valenzuela, "Computerized Cognitive Training in Cognitively Healthy Older Adults: A Systematic Review and Meta-Analysis of Effect Modifiers," *PLoS Medicine* 11, no. 11 (2014): e1001756; M. E. Kelly, D. Loughrey, B. A. Lawlor, I. H. Robertson, C. Walsh, and S Brennan, "The Impact of Cognitive Training and Mental Stimulation on Cognitive and Everyday Functioning of Healthy Older Adults: A Systematic Review and Meta-Analysis," *Aging Research Reviews* 15 (2014): 28–43.

19. K. Ball, D. B. Berch, K. F. Helmers, J. B. Jobe, M. D. Leveck, M. Marsiske, et al., "Effects of Cognitive Training Interventions with Older Adults: A Randomized Controlled Trial," *JAMA* 288, no. 18 (2002): 2271–2281.

20. G. W. Rebok, K. Ball, L. T. Guey, R. N. Jones, H. Y. Kim, J. W. King, et al., "Ten-Year Effects of the Advanced Cognitive Training for Independent and Vital Elderly Cognitive Training Trial on Cognition and Everyday Functioning in Older Adults," *Journal of the American Geriatrics Society* 62, no. 1 (2014): 16–24.

21. K. Ball, J. D. Edwards, L. A. Ross, and G. McGwin, Jr., "Cognitive Training Decreases Motor Vehicle Collision Involvement of Older Drivers," *Journal of the American Geriatrics Society* 58, no. 11 (2010): 2107–2113.

22. J. Mishra, E. de Villers-Sidani, M. Merzenich, and A. Gazzaley, "Adaptive Training Diminishes Distractibility in Aging across Species," *Neuron* 84, no. 5 (2014): 1091–1103.

23. J. Au, E. Sheehan, N. Tsai, G. J. Duncan, M. Buschkuehl, and S. M. Jaeggi, "Improving Fluid Intelligence with Training on Working Memory: A Meta-Analysis," *Psychonomic Bulletin and Review* (2014): 1–12; T. Klingberg, "Training and Plasticity of Working Memory," *Trends in Cognitive Sciences* 14, no. 7 (2010): 317–324; Y. Brehmer, H. Westerberg, and L. Bäckman, "Working-Memory Training in Younger and Older Adults: Training Gains, Transfer, and Maintenance," *Frontiers in Human Neuroscience*, 6 (2012): 63; J. Karbach and J. Kray, "How Useful Is Executive Control Training? Age Differences in Near and Far Transfer of Task-Switching Training," *Developmental Science*, 12, no. 6 (2009): 978–990; M. Lussier, C. Gagnon, and L. Bherer, "An Investigation of Response and Stimulus Modality Transfer Effects After Dual-Task Training in Younger and Older," *Frontiers in Human Neuroscience*, 6 (2012): 129.

24. A. P. Goldin, M. J. Hermida, D. E. Shalom, M. E. Costa, M. Lopez-Rosenfeld, M. S. Segretin, et al., "Far Transfer to Language and Math of a Short Software-Based Gaming Intervention," *Proceedings of the National Academy of Sciences* 111, no. 17 (2014): 6443–6448.

25. A. M. Owen, A. Hampshire, J. A. Grahn, R. Stenton, S. Dajani, A. S. Burns, et al., "Putting Brain Training to the Test," *Nature*, 465, no. 7299 (2010): 775–778.

26. W. Boot, D. Simons, C. Stothart, and C. Stutts, "The Pervasive Problem with Placebos in Psychology: Why Active Control Groups Are Not Sufficient to Rule Out Placebo Effects," *Perspectives on Psychological Science* 8, no. 4 (2013): 445–454.

27. C. S. Green and D. Bavelier, "Action Video Game Modifies Visual Selective Attention," *Nature* 423, no. 6939 (2003): 534–537.

28. Green and Bavelier, "Action Video Game."

29. C. S. Green and D. Bavelier, "Learning, Attentional Control, and Action Video Games," *Current Biology* 22, no. 6 (2012): R197–R206.

30. D. Bavelier, C. S. Green, A. Pouget, and P. Schrater, "Brain Plasticity through the Life Span: Learning to Learn and Action Video Games," *Neuroscience* 35 (2012): 391–416.

31. J. Mishra, M. Zinni, D. Bavelier, and S. A. Hillyard, "Neural Basis of Superior Performance of Action Videogame Players in an Attention-Demanding Task," *Journal of Neuroscience* 31, no. 3 (2011): 992–998.

32. D. Bavelier, R. L. Achtman, M. Mani, and J. Föcker, "Neural Bases of Selective Attention in Action Video Game Players," *Vision Research* 61 (2012): 132–143.

33. *NeuroRacer* game development team: Eric Johnston, Dmitri Ellingson, Matt Omernick from LucasArts, and game designer Noah Fahlstein.

34. J. A. Anguera, J. Boccanfuso, J. L. Rintoul, O. Al-Hashimi, F. Faraji, J. Janowich, et al., "Video Game Training Enhances Cognitive Control in Older Adults," *Nature* 501, no. 7465 (2013): 97–101.

35. A. M. Kueider, J. M. Parisi, A. L. Gross, and G. W. Rebok, "Computerized Cognitive Training with Older Adults: A Systematic Review," *PLoS ONE* 7, no. 7 (2012): e40588.

36. A. J. Latham, L. L. Patston, and L. J. Tippett, "The Virtual Brain: 30 Years of Video-Game Play and Cognitive Abilities," *Frontiers in Psychology* 4 (2013): 629.

37. S. Prot, K. A. McDonald, C. A. Anderson, and D. A. Gentile, "Video Games: Good, Bad, or Other?" *Pediatric Clinics of North America* 59, no. 3 (2012): 647–658.

38. D. A. Gentile, E. L. Swing, C. G. Lim, and A. Khoo, "Video Game Playing, Attention Problems, and Impulsiveness: Evidence of Bidirectional Causality," *Psychology of Popular Media Culture* 1, no. 1 (2012): 62.

39. Prot et al., "Video Games."

40. M. G. Berman, E. Kross, K. M. Krpan, M. K. Askren, A. Burson, P. J. Deldin, et al., "Interacting with Nature Improves Cognition and Affect for Individuals with Depression," *Journal of Affective Disorders* 140, no. 3 (2012): 300–305; A. F. Taylor and F. E. Kuo, "Children with Attention Deficits Concentrate Better after Walk in the Park," *Journal of Attention Disorders* 12, no. 5 (2009): 402–409.

41. S. Kaplan, "The Restorative Benefits of Nature: Toward an Integrative Framework," *Journal of Environmental Psychology* 15, no. 3 (1995): 169–182.

42. J. Persson, K. M. Welsh, J. Jonides, and P. A. Reuter-Lorenz, "Cognitive Fatigue of Executive Processes: Interaction between Interference Resolution Tasks," *Neuropsychologia* 45, no. 7 (2007): 1571–1579; M. Muraven and R. F. Baumeister, "Self-Regulation and Depletion of Limited Resources: Does Self-Control Resemble a Muscle?" *Psychological Bulletin* 126, no. 2 (2000): 247.

43. S. Kaplan, "The Restorative Environment: Nature and Human Experience," in *The Role of Horticulture in Human Well Being and Social Development*, ed. D. Relf (Portland, OR: Timber Press, 1992), 134–142.

44. R. S. Ulrich, "Aesthetic and Affective Response to Natural Environment," in *Behavior and the Natural Environment*, ed. I. Altman and J. F. Wohlwill (New York, Plenum Press, 1983), 85–125.

45. D. G. Pearson and T. Craig, "The Great Outdoors? Exploring the Mental Health Benefits of Natural Environments," *Frontiers in Psychology* 5 (2014): 1178.

46. B. Maher, "Poll Results: Look Who's Doping," *Nature* 452 (2008): 674–675.

47. S. E. McCabe, J. R. Knight, C. J. Teter, and H. Wechsler, "Non-medical Use of Prescription Stimulants among US College Students: Prevalence and Correlates from a National Survey," *Addiction* 100, no. 1 (2005): 96–106.

48. R. C. Spencer, D. M. Devilbiss, and C. W. Berridge, "The Cognition-Enhancing Effects of Psychostimulants Involve Direct Action in the Prefrontal Cortex," *Biological Psychiatry* 77, no. 11 (2015): 940–950.

49. C. Advokat, "What Are the Cognitive Effects of Stimulant Medications? Emphasis on Adults with Attention-Deficit/Hyperactivity Disorder (ADHD)," *Neuroscience and Biobehavioral Reviews* 34, no. 8 (2010): 1256–1266; L. C. Bidwell, F. J. McClernon, and S. H. Kollins, "Cognitive Enhancers for the Treatment of ADHD," *Pharmacology Biochemistry and Behavior* 99, no. 2 (2011): 262–274.

50. R. C. Malenka, E. J. Nestler, and S. E. Hyman, "Higher Cognitive Function and Behavioral Control," chapter 13 in *Molecular Neuropharmacology: A Foundation for Clinical Neuroscience*, 2nd ed., ed. A. Sydor and R. Y. Brown (New York: McGraw-Hill Medical, 2009), 318.

51. Advokat, "What Are the Cognitive Effects of Stimulant Medications?"

52. B. Sahakian and S. Morein-Zamir, "Professor's Little Helper," *Nature* 450, no. 7173 (2007): 1157–1159; A. G. Franke, C. Bagusat, S. Rust, A. Engel, and K. Lieb, "Substances Used and Prevalence Rates of Pharmacological Cognitive Enhancement among Healthy Subjects," *European Archives of Psychiatry and Clinical Neuroscience* 264, no. 1 (2014): 83–90.

53. A. M. Kelley, C. M. Webb, J. R. Athy, S. Ley, and S. Gaydos, "Cognition Enhancement by Modafinil: A Meta-Analysis," *Aviation, Space, and Environmental Medicine* 83, no. 7 (2012): 685–690; M. J. Minzenberg and C. S. Carter, "Modafinil: A Review of Neurochemical Actions and Effects on Cognition," *Neuropsychopharmacology* 33, no. 7 (2008): 1477–1502; D. Repantis, P. Schlattmann, O. Laisney, and I. Heuser, "Modafinil and Methylphenidate for Neuroenhancement in Healthy Individuals: A Systematic Review," *Pharmacological Research* 62, no. 3 (2010): 187–206.

54. D. C. Turner, T. W. Robbins, L. Clark, A. R. Aron, J. Dowson, and B. J. Sahakian, "Cognitive Enhancing Effects of Modafinil in Healthy Volunteers," *Psychopharmacology* 165, no. 3 (2003): 260–269; D. C. Randall, A. Viswanath, P. Bharania, S. M. Elsabagh, D. E. Hartley, J. M. Shneerson, and S. E. File, "Does Modafinil Enhance Cognitive Performance in Young Volunteers Who Are Not Sleep-Deprived?" *Journal of Clinical Psychopharmacology* 25, no. 2 (2005): 175–179.

55. M. J. Farah, J. Illes, R. Cook-Deegan, H. Gardner, E. Kandel, P. King, et al., "Neurocognitive Enhancement: What Can We Do and What Should We Do?" *Nature Reviews Neuroscience* 5, no. 5 (2004): 421–425; B. J. Sahakian and S. Morein-Zamir, "Neuroethical Issues in Cognitive Enhancement," *Journal of Psychopharmacology* 25, no. 2 (2011): 197–204.

56. K. T. Khaw, N. Wareham, S. Bingham, A. Welch, R. Luben, and N. Day, "Combined Impact of Health Behaviors and Mortality in Men and Women: The EPIC-Norfolk Prospective Population Study," *Obstetrical and Gynecological Survey* 63 (2008): 376–377.

57. A. J. Daley, "Exercise Therapy and Mental Health in Clinical Populations: Is Exercise Therapy a Worthwhile Intervention?" *Advances in Psychiatric Treatment* 8 (2002): 262–270; R. Walsh, "Lifestyle and Mental Health," *American Psychologist* 66, no. 7 (2011): 579.

58. C. W. Cotman and N. C. Berchtold, "Exercise: A Behavioral Intervention to Enhance Brain Health and Plasticity," *Neuroscience* 25 (2002): 295–301; K. I. Erickson, M. W. Voss, R. S. Prakash, C. Basak, A. Szabo, L. Chaddock, et al., "Exercise Training Increases Size of Hippocampus and Improves Memory," *Proceedings of the National Academy of Sciences* 108, no. 7 (2011): 3017–3022; K. I. Erickson, C. A. Raji, O. L. Lopez, J. T. Becker, C. Rosano, A. B. Newman, et al., "Physical Activity Predicts Gray Matter Volume in Late Adulthood: The Cardiovascular Health Study," *Neurology* 75, no. 16 (2010): 1415–1422; M. W. Voss, R. S. Prakash, K. I. Erickson, C. Basak, L. Chaddock, J. S. Kim, et al., "Plasticity of Brain Networks in a Randomized Intervention Trial of Exercise Training in Older Adults," *Frontiers in Aging Neuroscience* 2 (2010); K. Fabel and G. Kempermann, "Physical Activity and the Regulation of Neurogenesis in the Adult and Aging Brain," *Neuromolecular Medicine* 10, no. 2 (2008): 59–66; S. J. Colcombe, K. I. Erickson, P. E. Scalf, J. S. Kim, R. Prakash, E. McAuley, et al., "Aerobic Exercise Training

Increases Brain Volume in Aging Humans," *Journals of Gerontology Series A: Biological Sciences and Medical Sciences* 61, no. 11 (2006): 1166–1170.

59. S. B. Hindin and E. M. Zelinski, "Extended Practice and Aerobic Exercise Interventions Benefit Untrained Cognitive Outcomes in Older Adults: A Meta-Analysis," *Journal of the American Geriatric Society* 60 (2012): 136–141; B. A. Sibley and J. L. Etnier, "The Relationship between Physical Activity and Cognition in Children: A Meta-Analysis," *Pediatric Exercise Science* 15, no. 3 (2003): 243–256.

60. M. B. Pontifex, L. B. Raine, C. R. Johnson, L. Chaddock, M. W. Voss, N. J. Cohen, et al., "Cardiorespiratory Fitness and the Flexible Modulation of Cognitive Control in Preadolescent Children," *Journal of Cognitive Neuroscience* 23, no. 6 (2011): 1332–1345; M. R. Scudder, K. Lambourne, E. S. Drollette, S. D. Herrmann, R. A. Washburn, J. E. Donnelly, and C. H. Hillman, "Aerobic Capacity and Cognitive Control in Elementary School-Age Children," *Medicine and Science in Sports and Exercise* 46 (2014): 1025–1035.

61. M. B. Pontifex, L. B. Raine, C. R. Johnson, L. Chaddock, M. W. Voss, N. J. Cohen, et al., "Cardiorespiratory Fitness and the Flexible Modulation of Cognitive Control in Preadolescent Children," *Journal of Cognitive Neuroscience* 23, no. 6 (2011): 1332–1345.

62. J. R. Themanson, M. B. Pontifex, and C. H. Hillman, "Fitness and Action Monitoring: Evidence for Improved Cognitive Flexibility in Young Adults," *Neuroscience* 157, no. 2 (2008): 319–328.

63. D. Stavrinos, K. W. Byington, and D. C. Schwebel, "Effect of Cell Phone Distraction on Pediatric Pedestrian Injury Risk," *Pediatrics* 123, no. 2 (2009): e179–e185.

64. L. Chaddock, M. B. Neider, A. Lutz, C. H. Hillman, and A. F. Kramer, "Role of Childhood Aerobic Fitness in Successful Street Crossing," *Medicine and Science in Sports and Exercise* 44 (2012): 749–753.

65. J. R. Best, "Effects of Physical Activity on Children's Executive Function: Contributions of Experimental Research on Aerobic Exercise," *Developmental Review* 30, no. 4 (2010): 331–351; C. H. Hillman, M. B. Pontifex, D. M. Castelli, N. A. Khan, L. B. Raine, M. R. Scudder, et al., "Effects of the FITKids Randomized Controlled Trial on Executive Control and Brain Function," *Pediatrics* 134, no. 4 (2014): e1063–e1071.

66. Y. K. Chang, S. Liu, H. H. Yu, and Y. H. Lee, "Effect of Acute Exercise on Executive Function in Children with Attention Deficit Hyperactivity Disorder," *Archives of Clinical Neuropsychology* 27, no. 2 (2012): 225–237; C. H. Hillman, M. B. Pontifex, L. B. Raine, D. M. Castelli, E. E. Hall, and A. F. Kramer, "The Effect of Acute Treadmill Walking on Cognitive Control and Academic Achievement in Preadolescent Children," *Neuroscience* 159, no. 3 (2009): 1044–1054.

67. J. J. Ratey and J. E. Loehr, "The Positive Impact of Physical Activity on Cognition during Adulthood: A Review of Underlying Mechanisms, Evidence, and Recommendations," *Reviews in the Neurosciences* 22, no. 2 (2011): 171–185; C. L. Hogan, J. Mata, and L. L. Carstensen, "Exercise Holds Immediate Benefits for Affect and Cognition in Younger and Older Adults," *Psychology and Aging* 28, no. 2 (2013): 587–594; J. Barenberg, T. Berse, and S. Dutke, "Executive Functions in Learning Processes: Do They Benefit from Physical Activity?" *Educational Research Review* 6, no. 3 (2011): 208–222.

68. S. Colcombe and A. F. Kramer, "Fitness Effects on the Cognitive Function of Older Adults: A Meta-Analytic Study," *Psychological Science* 14, no. 2 (2003): 125–130.

69. E. McAuley, S. P. Mullen, and C. H. Hillman, "Physical Activity, Cardiorespiratory Fitness, and Cognition across the Lifespan," in *Social Neuroscience and Public Health*, ed. P. A. Hall (New York: Springer, 2013), 235–252.

70. J. E. Karr, C. N. Areshenkoff, P. Rast, and M. A. Garcia-Barrera, "An Empirical Comparison of the Therapeutic Benefits of Physical Exercise and Cognitive Training on the Executive Functions of Older Adults: A Meta-Analysis of Controlled Trials," *Neuropsychology* 28, no. 6 (2014): 829–845; L. Bherer, K. I. Erickson, and T. Liu-Ambrose, "A Review of the Effects of Physical Activity and Exercise on Cognitive and Brain Functions in Older Adults," *Journal of Aging Research* (2013).

71. S. J. Colcombe, A. F. Kramer, K. I. Erickson, P. Scalf, E. McAuley, N. J. Cohen, et al., "Cardiovascular Fitness, Cortical Plasticity, and Aging," *Proceedings of the National Academy of Sciences of the United States of America* 101, no. 9 (2004): 3316–3321; C. L. Davis, P. D. Tomporowski, J. E. McDowell, B. P. Austin, P. H. Miller, N. E. Yanasak, et al., "Exercise Improves Executive Function and Achievement and Alters Brain Activation in Overweight Children: A Randomized, Controlled Trial," *Health Psychology* 30, no. 1 (2011): 91.

72. H. van Praag, "Exercise and the Brain: Something to Chew On," *Trends in Neurosciences* 32, no. 5 (2009): 283–290; P. D. Bamidis, A. B. Vivas, C. Styliadis, C. Frantzidis, M. Klados, W. Schlee, et al., "A Review of Physical and Cognitive Interventions in Aging," *Neuroscience and Biobehavioral Reviews* 44 (2014): 206–220.

73. B. Zoefel, R. J. Huster, and C. S. Herrmann, "Neurofeedback Training of the Upper Alpha Frequency Band in EEG Improves Cognitive Performance," *Neuroimage* 54, no. 2 (2011): 1427–1431.

74. T. Ros, J. Théberge, P. A. Frewen, R. Kluetsch, M. Densmore, V. D. Calhoun, and R. A. Lanius, "Mind Over Chatter: Plastic Up-Regulation of the fMRI Salience Network Directly after EEG Neurofeedback," *Neuroimage* 65 (2013): 324–335.

75. J. H. Gruzelier, "EEG-Neurofeedback for Optimising Performance. I: A Review of Cognitive and Affective Outcome in Healthy Participants," *Neuroscience and Biobehavioral Reviews* 44 (2014): 124–141; D. J. Vernon, "Can Neurofeedback Training Enhance Performance? An Evaluation of the Evidence with Implications for Future Research," *Applied Psychophysiology and Biofeedback* 30, no. 4 (2005): 347–364.

76. S. K. Loo and R. A. Barkley, "Clinical Utility of EEG in Attention Deficit Hyperactivity Disorder," *Applied Neuropsychology* 12, no. 2 (2005): 64–76; M. Arns, H. Heinrich, and U. Strehl, "Evaluation of Neurofeedback in ADHD: The Long and Winding Road," *Biological Psychology* (2013): 1–8; D. C. Hammond, "Neurofeedback with Anxiety and Affective Disorders," *Child and Adolescent Psychiatric Clinics of North America* 14, no. 1 (2005): 105–123; M. E. J. Kouijzera, J. M. H. de Moor, B. J. L. Gerrits, J. K. Buitelaar, and Hein T. van Schie, "Long-Term Effects of Neurofeedback Treatment in Autism," *Research in Autism Spectrum Disorders* 3, no. 2 (2009): 496–501; F. Dehghani-Arani, R. Rostami, and H. Nadali, "Neurofeedback Training for Opiate Addiction: Improvement of Mental Health and Craving," *Applied Psychophysiology and Biofeedback* 38, no. 2 (2013): 133–141.

77. J. R. Wang and S. Hsieh, "Neurofeedback Training Improves Attention and Working Memory Performance," *Clinical Neurophysiology* 124, no. 12 (2013): 2406–2420.

78. J. Ghaziri, A. Tucholka, V. Larue, M. Blanchette-Sylvestre, G. Reyburn, G. Gilbert, et al. "Neurofeedback Training Induces Changes in White and Gray Matter," *Clinical EEG and Neuroscience* 44 (2013): 265–272.

79. D. Fox, "Neuroscience: Brain Buzz," *Nature News* 472, no. 7342 (2011): 156–159; H. L. Filmer, P. E. Dux, and J. B. Mattingley, "Applications of Transcranial Direct Current Stimulation for Understanding Brain Function," *Trends in Neurosciences* 37, no. 12 (2014): 742–753.

80. M. A. Nitsche and W. Paulus, "Excitability Changes Induced in the Human Motor Cortex by Weak Transcranial Direct Current Stimulation," *Journal of Physiology* 527, no. 3 (2000): 633–639.

81. R. Lindenberg, V. Renga, L. L. Zhu, D. Nair, and G. Schlaug, "Bihemispheric Brain Stimulation Facilitates Motor Recovery in Chronic Stroke Patients," *Neurology* 75, no. 24 (2010): 2176–2184; D. H. Benninger, M. Lomarev, G. Lopez, E. M. Wassermann, X. Li, E. Considine, and M. Hallett, "Transcranial Direct Current Stimulation for the Treatment of Parkinson's Disease," *Journal of Neurology, Neurosurgery, and Psychiatry* 81, no. 10 (2010): 1105–1111; A. R. Brunoni, L. Valiengo, A. Baccaro, T. A. Zanao, J. F. de Oliveira, A. Goulart, et al., "The Sertraline vs. Electrical Current Therapy for Treating Depression Clinical Study: Results from a Factorial, Randomized, Controlled Trial," *JAMA Psychiatry* 70, no. 4 (2013): 383–391.

82. M. F. Kuo and M. A. Nitsche, "Effects of Transcranial Electrical Stimulation on Cognition," *Clinical EEG and Neuroscience* 43, no. 3 (2012): 192–199.

83. B. A. Coffman, M. C. Trumbo, and V. P. Clark, "Enhancement of Object Detection with Transcranial Direct Current Stimulation Is Associated with Increased Attention," *BMC Neuroscience* 13, no. 1 (2012): 108; A. R. Brunoni and M.-A. Vanderhasselt, "Working Memory Improvement with Non-invasive Brain Stimulation of the Dorsolateral Prefrontal Cortex: A Systematic Review and Meta-Analysis," *Brain and Cognition* 86 (2014): 1–9; T. Strobach, A. Soutschek, D. Antonenko, A. Flöel, and T. Schubert, "Modulation of Executive Control in Dual Tasks with Transcranial Direct Current Stimulation (tDCS)," *Neuropsychologia* 68C (2014): 8–20.

84. J. Nelson, R. McKinley, E. Golob, J. Warm, and R. Parasuraman, "Enhancing Vigilance in Operators with Prefrontal Cortex Transcranial Direct Current Stimulation (tDCS)," *NeuroImage* 85 (2014): 909–917.

85. Nelson et al., "Enhancing Vigilance in Operators," 909.

86. W. Y. Hsu, T. P. Zanto, and A. Gazzaley, "Anodal Transcranial Direct Current Stimulation of Dorsolateral Prefrontal Cortex Enhances Multitasking Performance," paper presented at the Bay Area Memory Meeting, San Francisco, CA, 2013.

87. D. Reato, A. Rahman, M. Bikson, and L. C. Parra, "Effects of Weak Transcranial Alternating Current Stimulation on Brain Activity: A Review of Known Mechanisms from Animal Studies," *Frontiers in Human Neuroscience* 7 (2013): 687–687.

88. N. Jaušovec and K. Jaušovec, "Increasing Working Memory Capacity with Theta Transcranial Alternating Current Stimulation (tACS)," *Biological Psychology* 96 (2014): 42–47.

89. A. R. Brunoni, J. Amadera, B. Berbel, M. S. Volz, B. G. Rizzerio, and F. Fregni, "A Systematic Review on Reporting and Assessment of Adverse Effects Associated with Transcranial Direct Current Stimulation," *International Journal of Neuropsychopharmacology* 14, no. 8 (2011): 1133–1145.

90. T. Iuculano and R. Cohen Kadosh, "The Mental Cost of Cognitive Enhancement," *Journal of Neuroscience* 33, no. 10 (2013): 4482–4486.

91. T. Pustovrh, "The Neuroenhancement of Healthy Individuals Using tDCS: Some Ethical, Legal, and Societal Aspects," *Interdisciplinary Description of Complex Systems* 12, no. 4 (2014): 270–279.

92. J. Mishra and A. Gazzaley, "Closed-Loop Rehabilitation of Age-Related Cognitive Disorders," *Seminars in Neurology* 34 (2014): 584–590.

93. P. D. Bamidis, A. B. Vivas, C. Styliadis, C. Frantzidis, M. Klados, W. Schlee, et al., "A Review of Physical and Cognitive Interventions in Aging," *Neuroscience and Biobehavioral Reviews* 44 (2014): 206–220.

CHAPTER 11: MODIFYING BEHAVIOR

1. C. D. Fisher, "Effects of External and Internal Interruptions on Boredom at Work: Two Studies," *Journal of Organizational Behavior* 19, no. 5 (1998): 503–522.

2. J. Stern, "No Internet for One Year: Tech Writer Tries Life Offline," ABC News, July 19, 2012, http://abcnews.go.com/Technology/internet-year-technology-writer-paul-miller-life-offline/story ?id=16812425; http://99daysoffreedom.com/; W. Kagan, "Daring to Be Silent," *Chronogram— Hudson Valley News*, January 1, 2015, http://www.chronogram.com/hudsonvalley/daring-to-be-silent/Content?oid=2287720; W. L. Bjorklund, D. L. Rehling, P. S. Tompkins, and R. E. Strom, "The 'Unplugged' Project," *Communication Teacher* 26, no. 2 (2012): 92–95; http://national dayofunplugging.com/; A. Isaacson, "Learning to Let Go: First, Turn Off the Phone," *New York Times*, December 14, 2012, http://www.nytimes.com/2012/12/16/fashion/teaching-people-to -live-without-digital-devices.html.

3. An extensive examination of the literature found only two peer-reviewed journal articles that examine the effectiveness of digital detoxes and assess their efficacy: C. M. Paris, E. A. Berger, S. Rubin, and M. Casson, "Disconnected and Unplugged: Experiences of Technology Induced Anxieties and Tensions While Traveling," in *Information and Communication Technologies in Tourism 2015* (New York: Springer International, 2015), 803–816; S. Y. Schoenebeck, "Giving Up Twitter for Lent: How and Why We Take Breaks from Social Media," in *CHI '14: Proceedings of the SIGCHI Conference on Human Factors in Computing Systems* (New York: ACM Press, 2014), 773–782.

4. S. Turkle, *Alone Together: Why We Expect More from Technology and Less from Each Other* (New York: Basic Books, 2011); Y. T. Uhls, M. Michikyan, J. Morris, D. Garcia, G. W. Small, E. Zgourou, and P. M. Greenfield, "Five Days at Outdoor Education Camp without Screens Improves Preteen Skills with Nonverbal Emotion Cues," *Computers in Human Behavior*, 39 (2014): 387–392; A. K. Przybylski and N. Weinstein, "Can You Connect with Me Now? How the Presence of Mobile Communication Technology Influences Face-to-Face Conversation Quality," *Journal of Social and Personal Relationships* 30, no. 3 (2013): 237–246.

5. http://www.fcc.gov/guides/texting-while-driving.

6. http://www.itcanwaitsimulator.org/.

7. For up-to-date information and research, see http://nsc.org/handsfree; Governors Highway Safety Association, "Distracted Driving Laws" (February 2015), http://www.ghsa.org/html/stateinfo/laws/cellphone_laws.html.

8. Governors Highway Safety Association, "Distracted Driving: What Research Shows and What States Can Do" (2011), http://www.ghsa.org/html/files/pubs/sfdist11.pdf. Here is a sampling of recent studies that have examined the impact of handsfree driving: M. M. Haque and S. Washington, "The Impact of Mobile Phone Distraction on the Braking Behavior of Young Drivers: A Hazard-Based Duration Model," *Transportation Research Part C: Emerging Technologies* 50 (2015): 13–27; B. Metz, A. Landau, and V. Hargutt, "Frequency and Impact of Hands-Free Telephoning While Driving—Results from Naturalistic Driving Data," *Transportation Research Part F: Traffic Psychology and Behavior* 29 (2015): 1–13.

9. M. L. Aust, A. Eugensson, J. Ivarsson, and M. Petersson, *Thinking about Distraction—A Conceptual Framework for Assessing Driver-Vehicle On-Road Performance in Relation to Secondary Task Activity*, National Highway Transportation and Safety Administration Report, http://www-nrd .nhtsa.dot.gov/pdf/Esv/esv22/22ESV-000320.pdf; L. Nunes and M. A. Recarte, "Cognitive Demands of Hands-Free-Phone Conversation While Driving," *Transportation Research Part F: Traffic Psychology and Behavior* 5, no. 2 (2002): 133–144.

10. D. L. Strayer, J. Turrill, J. R. Coleman, E. V. Ortiz, and J. M. Cooper, "Measuring Cognitive Distraction in the Automobile II: Assessing In-Vehicle Voice-Based Interactive Technologies," *Accident Analysis and Prevention* 372 (2014): 379; D. Strayer, J. Cooper, J. Turrill, J. Coleman, N. Medeiros-Ward, and F. Biondi, *Measuring Cognitive Distraction in the Automobile* (2013), AAA Foundation for Traffic Safety, Washington, DC.

11. F. Manjoo, "Discovering Two Screens Aren't Better Than One," *New York Times*, March 19, 2014, http://www.nytimes.com/2014/03/20/technology/personaltech/surviving-and-thriving-in -a-one-monitor-world.html.

12. C. Thompson, "End the Tyranny of 24.7 Email," *New York Times*, August 28, 2014, http://www .nytimes.com/2014/08/29/opinion/end-the-tyranny-of-24-7-email.html.

13. K. Kushlev and E. W. Dunn, "Checking Email Less Frequently Reduces Stress," *Computers in Human Behavior* 43 (2015): 220–228.

14. D. J. Levitan, "Hit the Reset Button in Your Brain," *New York Times*, August 9, 2014, http:// www.nytimes.com/2014/08/10/opinion/sunday/hit-the-reset-button-in-your-brain.

15. B. Thornton, A. Faires, M. Robbins, and E. Rollins, "The Mere Presence of a Cell Phone May Be Distracting: Implications for Attention and Task Performance," *Social Psychology* 45 (2014): 479–488.

16. A. Mirelman, I. Maidan, H. Bernad-Elazari, F. Nieuwhof, M. Reelick, N. Giladi, and J. M. Hausdorff, "Increased Frontal Brain Activation during Walking While Dual Tasking: An fNIRS Study in Healthy Young Adults," *Journal of Neuroengineering and Rehabilitation* 11, no. 1 (2014): 85; R. Beurskens, I. Helmich, R. Rein, and O. Bock, "Age-Related Changes in Prefrontal Activity during Walking in Dual-Task Situations: A fNIRS Study," *International Journal of Psychophysiology* 92, no. 3 (2014): 122–128.

17. T. Lesiuk, "The Effect of Preferred Music on Mood and Performance in a High-Cognitive Demand Occupation," *Journal of Music Therapy* 47, no. 2 (2010): 137–154; L. A. Angel, D. J. Polzella, and G. C. Elvers, "Background Music and Cognitive Performance," *Perceptual and Motor Skills* 110, no. 3 (pt. 2) (2010): 1059–1064.

18. J. Blascovich, "Effects of Music on Cardiovascular Reactivity among Surgeons," *JAMA* 272, no. 11 (1994): 882–884; D. N. Moris and D. Linos, "Music Meets Surgery: Two Sides to the Art of 'Healing,'" *Surgical Endoscopy* 27, no. 3 (2013): 719–723; M. V. Thoma, M. Zemp, L. Kreien-bühl, D. Hofer, P. R. Schmidlin, T. Attin, et al., "Effects of Music Listening on Pre-treatment Anxiety and Stress Levels in a Dental Hygiene Recall Population," *International Journal of Behavioral Medicine* 22, no. 4 (2015): 498–505.

19. S. Ransdell and L. Gilroy, "The Effects of Background Music on Word Processed Writing," *Computers in Human Behavior* 17, no. 2 (2001): 141–148.

20. A. Ariga and A. Lleras, "Brief and Rare Mental 'Breaks' Keep You Focused: Deactivation and Reactivation of Task Goals Preempt Vigilance Decrements," *Cognition* 118, no. 3 (2011): 439–443.

21. C. Popovich and W. R. Staines, "Acute Aerobic Exercise Enhances Attentional Modulation of Somatosensory Event-Related Potentials during a Tactile Discrimination Task," *Behavioral Brain Research* 281 (2014): 267–275; M. T. Tine and A. G. Butler, "Acute Aerobic Exercise Impacts Selective Attention: An Exceptional Boost in Lower-Income Children," *Educational Psychology* 32, no. 7 (2012): 821–834.

22. J. Kaldenberg, J. Tribley, S. McClain, A. Karbasi, and J. Kaldenberg, "Tips for Computer Vision Syndrome Relief and Prevention," *Work* 39, no. 1 (2011): 85–87; Z. Yan, L. Hu, H. Chen, and F. Lu, "Computer Vision Syndrome: A Widely Spreading but Largely Unknown Epidemic among Computer Users," Computers in Human Behavior 24, no. 5 (2008): 2026–2042.

23. R. M. Daniel, "The Effects of the Natural Environment on Attention Restoration" (doctoral dissertation, Appalachian State University, 2014).

24. J. C. McVay and M. J. Kane, "Why Does Working Memory Capacity Predict Variation in Reading Comprehension? On the Influence of Mind Wandering and Executive Attention," *Journal of Experimental Psychology: General* 141, no. 2 (2012): 302–320; D. R. Thomson, D. Smilek, and D. Besner, "On the Asymmetric Effects of Mind-Wandering on Levels of Processing at Encoding and Retrieval," *Psychonomic Bulletin and Review* 21, no. 3 (2014): 728–733; J. W. Schooler, M. D. Mrazek, M. S. Franklin, B. Baird, B. W. Mooneyham, C. Zedelius, and J. M. Broadway, "The Middle Way: Finding the Balance between Mindfulness and Mind-Wandering," *Psychology of Learning and Motivation* 60 (2014): 1–33.

25. T. L. Signal, P. H. Gander, H. Anderson, and S. Brash, "Schedule Napping as a Countermeasure to Sleepiness in Air Traffic Controllers," *Journal of Sleep Research* 18, no. 1 (2009): 11–19; J. S. Ruggiero and N. S. Redeker, "Effects of Napping on Sleepiness and Sleep-Related Performance Deficits in Night-Shift Workers: A Systematic Review," *Biological Research for Nursing* 16, no. 2 (2013): 134–142; L. Moore, "High School Students' Perceived Alertness in Afternoon Classes Following a Short Post-Lunch Nap" (doctoral dissertation, Northwest Missouri State University, 2014); A. Gardner, "'Powernaps' May Boost Right-Brain Activity," *CNN.com*, September 25, 2013, http://www.cnn.com/2012/10/17/health/health-naps-brain/.

26. B. Ditzen and M. Heinrichs, "Psychobiology of Social Support: The Social Dimension of Stress Buffering," *Restorative Neurology and Neuroscience* 32, no. 1 (2014): 149–162.

27. G. S. Bains, L. S. Berk, N. Daher, E. Lohman, E. Schwab, J. Petrofsky, and P. Deshpande, "The Effect of Humor on Short-term Memory in Older Adults: A New Component for Whole-Person Wellness," *Advances in Mind—Body Medicine* 28, no. 2 (2013): 16–24.

28. C. Seiter, "Why You Need to Stop Thinking You Are Too Busy to Take Breaks: Inside the Science of Why Taking Breaks Can Make You Happier and More Focused and Productive. Still Think You're Too Important?" *Fast Company*, September 2, 2014, http://www.fastcompany.com /3034928/the-future-of-work/why-you-need-to-stop-thinking-you-are-too-busy-to-take-breaks.

29. C. Goldman, "This Is Your Brain on Jane Austen, and Stanford Researchers Are Taking Notes," *Stanford Report*, September 7, 2012, http://news.stanford.edu/news/2012/september/austen -reading-fmri-090712.html.

30. M. Goyal, S. Singh, E. M. Sibinga, N. F. Gould, A. Rowland-Seymour, R. Sharma, J. A. Haythornthwaite, et al., "Meditation Programs for Psychological Stress and Well-Being: A Systematic Review and Meta-Analysis," *JAMA Internal Medicine* 174, no. 3 (2014): 357–368.

31. J. Yin and R. K. Dishman, "The Effect of Tai Chi and Qigong Practice on Depression and Anxiety Symptoms: A Systematic Review and Meta-Regression Analysis of Randomized Controlled Trials," *Mental Health and Physical Activity* 7, no. 3 (2014): 135–146; P. M. Wayne, J. N. Walsh, R. E. Taylor-Piliae, R. E. Wells, K. V. Papp, N. J. Donovan, and G. Y. Yeh, "Effect of Tai Chi on Cognitive Performance in Older Adults: Systematic Review and Meta-Analysis," *Journal of the American Geriatrics Society* 62, no. 1 (2014): 25–39; C. W. Wang, C. H. Chan, R. T. Ho, J. S. Chan, S. M. Ng, and C. L. Chan, "Managing Stress and Anxiety through Qigong Exercise in Healthy Adults: A Systematic Review and Meta-Analysis of Randomized Controlled Trials," *BMC Complementary and Alternative Medicine* 14, no. 1 (2014): 8; F. Wang, E. K. O. Lee, T. Wu, H. Benson, G., Fricchione, W. Wang, and A. S. Yeung, "The Effects of Tai Chi on Depression, Anxiety, and Psychological Well-Being: A Systematic Review and Meta-Analysis," *International Journal of Behavioral Medicine* 21, no. 4 (2014): 605–617; Goyal et al., "Meditation Programs for Psychological Stress and Well-Being."

32. Przybylski and Weinstein, "Can You Connect with Me Now?"; S. Misra, L. Cheng, J. Genevie, and M. Yuan, "The iPhone Effect: The Quality of In-Person Social Interactions in the Presence of Mobile Devices," *Environment and Behavior* 48, no. 2 (2016): 275–298; Thornton et al., "The Mere Presence of a Cell Phone May Be Distracting."

33. C. Chambliss, E. Short, J. Hopkins-DeSantis, H. Putnam, B. Martin, M. Millington, et al., "Young Adults' Experience of Mobile Device Disruption of Proximate Relationships," *International Journal of Virtual Worlds and Human Computer Interaction* 3 (2015): 2368–6103.

34. Take the nomophobia test at http://antinomophobe.activemobi.com/.

35. C. Friedersdorf, "Arianna Huffington Implores You: Stop Taking Bright Devices into the Bedroom," *Atlantic*, June 27, 2013, http://www.theatlantic.com/technology/archive/2013/06/arianna -huffington-implores-you-stop-taking-bright-devices-into-the-bedroom/277269/.

36. C. D. Watkins, P. J. Fraccaro, F. G. Smith, J. Vukovic, D. R. Feinberg, L. M. DeBruine, and B. C. Jones, "Taller Men Are Less Sensitive to Cues of Dominance in Other Men," *Behavioral Ecology* 21, no. 5 (2010): 943–947.

37. M. E. Harrison, M. L. Norris, N. Obeid, M. Fu, H. Weinstangel, and M. Sampson, "Systematic Review of the Effects of Family Meal Frequency on Psychosocial Outcomes in Youth," *Canadian Family Physician* 61, no. 2 (2015): e96–e106.

38. M. P. Judge, X. Cong, O. Harel, A. B. Courville, and C. J. Lammi-Keefe, "Maternal Consumption of a DHA-Containing Functional Food Benefits Infant Sleep Patterning: An Early Neurodevelopmental Measure," *Early Human Development* 88, no. 7 (2012): 531–537.

39. M. Wood, "Bedtime Technology for a Better Night's Sleep," *New York Times*, December 24, 2014, http://www.nytimes.com/2014/12/25/technology/personaltech/bedroom-technology-for-a-better-nights-sleep.html.

40. See http://sleepfoundation.org/bedroom/see.php for the National Sleep Foundation recommendations.

41. Mayo Clinic, "Are Smartphones Disrupting Your Sleep? Mayo Clinic Study Examines the Question," January 3, 2013, http://newsnetwork.mayoclinic.org/discussion/are-smartphones-disrupting-your-sleep-mayo-clinic-study-examines-the-question-238cef/?_ga=1.215025045.366396735.1415659551.

42. http://justgetflux.com.

43. R. Pinkham, "80% of Smartphone Users Check Their Phones before Brushing Their Teeth … and Other Hot Topics," April 5, 2013, http://blogs.constantcontact.com/smartphone-usage-statistics/.

44. C. F. Wang, Y. L. Sun, and H. X. Zang, "Music Therapy Improves Sleep Quality in Acute and Chronic Sleep Disorders: A Meta-Analysis of 10 Randomized Studies," *International Journal of Nursing Studies* 51, no. 1 (2014): 51–62.

45. G. de Niet, B. Tiemens, B. Lendemeijer, and G. Hutschemaekers, "Music-Assisted Relaxation to Improve Sleep Quality: Meta-Analysis," *Journal of Advanced Nursing* 65, no. 7 (2009): 1356–1364.

46. I. W. Saxvig, A. Wilhelmsen-Langeland, S. Pallesen, Ø. Vedaa, I. H. Nordhus, and B. Bjorvatn, "A Randomized Controlled Trial with Bright Light and Melatonin for Delayed Sleep Phase Disorder: Effects on Subjective and Objective Sleep," *Chronobiology International* 31, no. 1 (2014): 72–86; K. C. Smolders and Y. A. de Kort, "Bright Light and Mental Fatigue: Effects on Alertness, Vitality, Performance, and Physiological Arousal," *Journal of Environmental Psychology* 39 (2014): 77–91.

Index

Page numbers followed by an "f" indicate a figure.